A TREATISE ON STEEL:

COMPRISING

ITS THEORY, METALLURGY, PROPERTIES, PRACTICAL WORKING, AND USE.

BY

M. H. C. LANDRIN, Jr.,

CIVIL ENGINEER.

TRANSLATED FROM THE FRENCH, WITH NOTES,

By A. A. FESQUET,

CHEMIST AND ENGINEER.

WITH AN

APPENDIX

ON THE

BESSEMER AND THE MARTIN PROCESSES FOR MANUFACTURING STEEL,

FROM THE

REPORT OF ABRAM S. HEWITT, UNITED STATES COMMISSIONER TO THE UNIVERSAL EXPOSITION, PARIS, 1867.

PHILADELPHIA:
HENRY CAREY BAIRD,
INDUSTRIAL PUBLISHER,
406 Walnut Street.
LONDON:
TRÜBNER & CO.,
60 Paternoster Row.
1868.

PHILADELPHIA:
COLLINS, PRINTER, 705 JAYNE STREET.

TRANSLATOR'S PREFACE.

Not many years ago, when the uses of steel were confined to the manufacture of tools, cutting instruments, etc., there were few qualities of steel; and cast steel, tilted steel, and some German steels especially employed for the manufacture of files, scythes, etc., were the only compounds of iron and carbon known under the name of steel.

The enormous progress made recently in iron metallurgy and in metallic constructions, has prompted the employment in large quantities of new kinds of steel, endowed with properties of resistance and flexibility hitherto unknown and unsuspected, and at a cheapness of cost which has allowed their use for rails, large pieces of machinery, railroad tires, plates for boilers, ships, bridges, etc.

These improvements are due to the energy of some inventors, and to their complete knowledge of the nature and composition of the metal with which they were working. The scientific principles by which their efforts were directed were the same, the methods to arrive at the result were different.

These new steels, whose properties were so different from those hitherto known, made many persons inquire if they were really steels, or some particular kinds of iron. This depends on definitions: we shall call them steels, if we continue to consider as—

Iron, the fibrous or granular metal known under that name, combined with a trace of carbon (all commercial irons contain carbon), and which cannot be hardened by the hardening process.

Steel, the combination of iron with an average of one per cent. of carbon, which can be hardened and melted.

Pig iron, the combination of iron with three

to five per cent. of carbon, which can be melted, and possesses scarcely any malleability.

A very small change in the proportion of carbon will affect the properties of steel, which will be exemplified in the course of this work. Nevertheless, Mr. Landrin takes cast steel as a type, because of all known steels its properties and composition are the most constant.

Beginning with a history of steel, the Author next examines the various fuels employed in metallurgy, the substances which in the ore and the fuel are capable of influencing the qualities of iron and steel, the different ores in use, and then passes to the theory of the formation of steel, with citations from a work on steel by Réaumur, published in 1722. This ancient work may be read with advantage even at the present day, as pointing to the futility of certain secrets and mixtures for making steel, by which many persons are yet deceived.

The theory of the formation of steel is followed by a method of quantitative analysis for iron, steel, or pig metal, by the metallurgy of natural, puddled, cast steel, and Wootz steel, special attention being given to the manufacture of pots for casting, and by the new processes known under the names of Chenot, Bessemer, Uchatius, etc.

After examining certain mixtures of steel with other metals, Mr. Landrin, Jr., fully explains the various operations by which steel is welded, hardened, and tempered; and finally describes some of the uses to which steel is applied, such as the manufacture of files, steel wire, steel plates, needles, and saws.

Within a small compass, Mr. Landrin gives an insight into the whole question of the manufacture of steel. Formerly, the steel manufacturer had only to buy some well-known mark of iron, cement and cast it; but the present state of steel industry, where the pig iron is sometimes run directly from the blast furnace into the converter, requires a knowledge of the ores and fuels employed for producing the raw metal.

To extend the knowledge of a metal so necessary as steel, has been the aim of the Author, and we hope our readers will find this book useful to them.

In order to convey an idea of the present state of steel industry, we have added some extracts from the valuable report made by Mr. Abram S. Hewitt, U. S. Commissioner to the Universal Exposition at Paris, 1867.

PHILADELPHIA, August 1, 1868.

CONTENTS.

The figures refer to the numbers of the paragraphs.

INTRODUCTION.

HISTORY OF STEEL.

PRELIMINARY OBSERVATIONS.

I. Heat.

II. Oxygen.

PART FIRST.

STEEL AND ITS THEORY.

Theory of Réaumur.

QUANTITATIVE ANALYSIS.

PART SECOND.

METALLURGY OF STEEL.

PART THIRD.

WORKING OF STEEL.

PART FOURTH.

PROPERTIES OF STEEL AND ITS USES.

II. Files.

III. Steel Wire.

IV. Needles.

V. Steel Plate.

VI. Saws.

INTRODUCTION.

HISTORY OF STEEL.

1. The discovery of steel is lost in antiquity, and is mingled with that of iron; the first indications are found in Genesis. In order to write the history of the carburized metal, we should give that of iron; but this would carry us too far beyond the limits of this small treatise. The Hebrews have confounded under the name of ברזל (*barzel*), the two metals, the working of which, according to Moses, had been taught to men by Tubal-Cain[1], whose father lived 3130 years before Christ. Job says that the metal was extracted from an arenaceous ore, probably similar to that actually employed at Samakof, in Romalia.[2] We find in Deuteronomy that Og, King of Bashan, possessed an iron bedstead, which shows to what degree of perfection the art of forging had arrived.[3]

[1] And Zillah, she also bare Tubal-Cain, an instructor of every artificer in brass and iron. (Gen. iv. 22.)

[2] Job xxviii. 2.

[3] Deuteronomy iii. 11.

2. It is probable that, like what is now the usage in Asia and in Central Africa, the Hebrews had nothing which could be properly called a furnace; but that they made a hole in the ground, filled it pell-mell with ore and charcoal, and promoted the combustion with their breath first, and afterwards with fans made of large tree leaves. The first indications of a regular mode of working are to be found in Egypt, where the Jews, during their servitude, were obliged to work in the forges of that country.[1]

3. Already, at that time, steel and iron were known by the Chinese; their first chiefs had found iron mines in the territory of Leang-tcheou.[2] It is likely that the knowledge of the metal came from the west of Peking, because it is there where the Celestial empire began to be populated.

4. In Larcher's Chronology it is stated that iron was known only 1537 years before Christ, over 250 years before the Trojan war. In the age of Homer, the metallurgical working of iron and steel was much advanced and varied; the latter metal was polished, as we may infer from the epithets αἴθωνι, *shiny*, and πολιός, *white*,[3] applied by the poet in his

[1] But the Lord hath taken you, and brought you forth out of the iron furnace (כורברזל, *Courbarzel*), even out of Egypt. (Deuter. iv. 20.)

[2] Chu-King, cap. Yu-Cong.

[3] Il. iv. 485; vii. 473; xx. 372.

poems, in opposition to μέλας, *black*, which seems to indicate ductile iron, as it comes from the hands of the blacksmith. It was hardened, because in the Odyssey, Homer compares the noise of the inflamed branch driven into the eye of Polyphemus to that produced by a blacksmith, when dipping into cold water a saw or an axe for hardening—an operation from which is derived all the strength of iron.[1]

5. It is a mistake to believe that in the heroic age, bronze alone was employed in the manufacture of arms and agricultural implements; from Homer we learn that the points of darts were sometimes forged out of steel,[2] and this metal was also used to make tilling instruments and weapons for shepherds.[3]

6. It is true that the use of bronze had preceded that of steel, as shown by Hesiod,[4] and after him, by Lucretius;[5] but we must not conclude with Eustathius and several other commentators, that Homer had confounded the two metals under the generic name of χαλκός.

This error is very likely caused by the Grecian poet confounding all metal workers under the same

[1] Odyss. ix. 391. The word φαρμάσσειν used by Homer, is also employed by Greek dyers to express the action of dipping the cloth into the bath.

[2] Odyss. xix. 494.

[3] Il. xxiii. 832.

[4] Ερσα καὶ ἡμέραι, 150 and 151.

[5] V. 1286.

denomination, that of χαλκεύς, χαλκεύειν, which means the *working of metals*, whatsoever be their nature; and moreover the root is χαλκός, *bronze*. Suidas and Hisychius express the word χαλκεὺς, by a *worker in bronze, iron, or gold*.

7. According to Aristotle, in ancient times, bronze was manufactured, and vases were cast in the island of Æthalia (Elba), celebrated for the richness of its iron mines. It is more recently, and when the ores for bronze making were exhausted, that iron was discovered, and its extraction began.[1]

8. From a memorable passage of that book of Aristotle called *Concealed Marvels*, we learn that the Chalybes worked a ferruginous sand, after a simple washing, without any admixture: it was difficult then for them to obtain anything else but steel.[2] In some cases, the washing was more perfect, and a flux was added, which they called *pyrimac*.[3] Theophrastes, while confirming this assertion, says that the pyrimac and molar stones used were very fusible,

[1] De mirabil. auscult. expl., by J. Beckmann, cap. xcv.

[2] The Chalybes inhabited the southern coast of Pontus Euxinus, where they had become celebrated by their manufactures of steel. The Greeks called χάλυβς the metal they imported from that country: from thence the word went to the Romans, afterwards to Spain, and finally to Great Britain, where the term *chalybeate* has been kept and applied to various uses of iron and steel.

[3] De mir. ausc., cap. xlix. pp. 92, 95.

and helped the fusion of iron.[1] It is likely that the *pyrimac* or *pyromac* stone was the variety of silica called *Flint*[2] (*pierre à feu*, Fr., *feuerstein*, Ger.), and that the *molar* stone was simply a kind of limestone. This is perfectly well explained by Pliny: *igne cremato lapide cœmenta in tectis ligantur.*[3]

9. The operation was performed in a low furnace. Aristotle, in a passage which commentators explain in different ways, seems to say that the operation required two fires—probably one was for reducing, and the other for reheating. This explanation of the Greek text appears to be proper, since it is what occurs actually in the forges we have inherited from the ancients. Moreover, Pliny says so formally.[4]

10. The fuel used was mostly wood. Pliny indicates pine-wood as being used in the manufacture of iron and steel. It is certain that the ancients had cognizance of charcoal, because they employed it in several metallurgical processes; but nowhere is it said that it was used in the working of steel.

11. Bituminous coal (*lithanthrax*) was known and employed by blacksmiths[5] in the earliest ages of

[1] Περὶ λίθων, § 19.

[2] In dictionaries the words πυρίμακος or πυρόμακος are translated into *fire-stone.*

[3] Lib. xxxvi. cap. xxvi.

[4] Lib. xxxiv. cap. xiv.

[5] Theophrastus, cap. xxvii.

Greece and Rome; nevertheless, no indications can be found that it was employed largely in the metallurgy of iron. Such coal was found in Elea, on the road to Olympia, and in Liguria, where amber was to be obtained also.[1]

12. There is no doubt that pig metal or raw iron was known by the ancients. Aristotle shows this clearly and precisely. "Iron," says he, "is softened during the operation so as to become liquid, but soon it hardens again. This is the only way of making steel. A pasty scoria swims on top, while the separated iron falls to the bottom."[2] Nevertheless, it appears that it never came to the mind of metallurgists of those ages to turn to account this state of fluidity, and to run the metal into moulds. It is possible that the rapidity with which pig metal coagulates and hardens has been found to be an obstacle to its use.

Pliny had known also this state of liquefaction. "It is surprising," says he, "when the ore is reduced, to see the iron become as fluid as water, and be afterwards divided sponge like."[3] The word *spongiæ*, which, among German commentators,[4] has received numerous interpretations, is here evidently applied

[1] Theophrastus, cap. xxviii.

[2] Μετεωρολογικῶν, cap. vi.

[3] Commendatio de arte ferri conficiendi veterum, Haussmann, p. 41 and seq.

[4] Cap. viii. v. 9.

to the mass full of scoriæ, which, most generally, is divided in several pieces in order to be brought more conveniently under the hammer.

13. The art of extracting, refining, and working iron and steel was diffused from Egypt over all parts of Arabia. It was known of old, that the mountainous part of Palestine was rich in iron mines; thus in Deuteronomy, it was said with some appearance of truth to the Israelites, on the eve of entering that land, that its stones were iron.[1]

14. The Jews and Egyptians were not alone in knowing how to forge iron and steel, inasmuch as Og, King of Bashan, had a bedstead made of that metal.[2] At Babylon, in place of hydraulic cement, the stones of bridges were bound together by means of iron bars cemented with melted lead.[3] Instruments made of steel were multiplied very rapidly: they were used for cutting stones[4] and for committing murder.[5] Ductile iron was employed for a great many uses; even nails were made out of it;[6] and

[1] Deuter. viii. 9.

[2] Ibid. iii. 11.

[3] Herod. lib. i. 186. Thus were consolidated on their outer surface the walls of Pyrei (Thucydides, lib. i. cap. xcxiii.): they were made of large stones, fitting very closely, cut square, and united without cement, lime, or clay.

[4] Deuter. xxvii. 5.

[5] Numbers xxxv. 16.

[6] Judges iv. 21.

Sisera, Captain of Jabin's army, had nine hundred chariots of iron.[1]

15. At that epoch, locksmiths or blacksmiths[2] were important personages, whom kings took with them on their journeys, as we may see when Jehoiakim, King of Judah, left Jerusalem.[3]

16. We have not now to inquire what were the iron ores treated by the ancients; this important question will subsequently receive due attention. Besides the arenaceous magnetic ore which, we have said, was employed by the Hebrews and Greeks of old,[4] it is likely that the other ores used by ancient metallurgists were those whose external appearance indicated the most plainly the nature of the metal within, whose extraction was the least difficult and laborious, and whose fusion was easy.

17. However, in proportion as Roman civilization

[1] Judges iv. 13.

[2] מסגר (*Mashagour*). This word is translated by Buxtorf into *faber ferrarius, claustrarius.* This interpretation is conformable to the root סגר (*shagour*), which means *to lock, to inclose* (Gen. ii. 21; vii. 16; xix. 6; Exod. xiv. 3; Jos. vi. 1.

[3] 2 Kings xxiv. 13, 15; Jerem. xxiv. 1; xxix. 2.

[4] The iron and steel of Avelino, near Naples, is made out of a similar ore. We find also the same ore worked in Tyrol; also by the savages in Virginia; in the small furnaces of Heraclea, on the shores of the Black Sea, &c.

took the place of Grecian, so was the use of iron and steel extended. Before the historical era, and up to Herodotus, bronze had been quite exclusively the material from which weapons were made. Under the Romans iron and steel had been put to more uses, and took the place of copper in many cases.

The two greatest poets of antiquity, Homer and Virgil, who delineate so well the different epochs in which they lived, state in a striking manner by two verses, and with similar expressions, the preponderance of one metal over the other at different historical times. Homer, speaking of death, says: χάλκιον ὕπνον, sleep of bronze;[1] Virgil expresses the same idea by *ferreus somnus*, sleep of iron.[2]

18. The art of forging iron and steel infers the employment of the hammer and the anvil. Thus it is likely that these two simple implements were known by Tubal-Cain. Goguet asserts they are mentioned in Job.[3] The first notice of these tools is found in Isaiah,[4] when he says: "So the carpenter encouraged the goldsmith, and he that smooth-

[1] Il. v. 785. [2] Eneid. x. 745; xii. 309.

[3] Cap. xli. 15, 20 (Orig. des Lois, p. 172). This is apparently a mistake. I have vainly endeavored to find that passage in the Hebrew text. Mention is made of the use of hammer and spikes in the book of Judges (iv. 21).

[4] Isaiah xli. 7, פטש (*Pethist*). In chap. xliv. 12, the same prophet uses the expression מקבת (*Makebet*) for a peculiar ham-

eth with the hammer him that smote the anvil, saying, it is ready for the sodering; and he fastened it with nails, that it should not be moved."

19. It is probable that bellows were employed more recently. The word used in the metallurgy of the Hebrews to express the action of blowing, is נפח (*nophéh*).[1] Jeremiah has even made a name of that word, to indicate a blowing apparatus.[2] The Septante have translated by φυσητής, and the Vulgate by *sufflatorium.*

Homer is much more explicit in the description of these instruments. The bellows (φῦσαι) of Vulcan were movable; they revolved around a pivot, (ἔγρεψε). The anvil (ἄκμων)[3] could be taken off the stock[4] at will; anvils of several sizes were employed. The tongs (πυράγρα) derived their name from their use.[5]

20. However, at about the time when Moses was leading the Hebrews across the deserts of Arabia, the cities of Tyre and Sidon were founded by Egyptian colonies on the shores of the Mediterranean

mer, one side of which had a flat face, while the other side was pointed.

[1] Isaiah liv. 16; Ezekiel xxii. 20, 21.

[2] Jerem. vi. 29.

[3] Il. xviii. 476.

[4] 'Ακμόθεζον, made of ἄκμων, anvil, and of τίθημι, to put on, to place, to sit.

[5] From πῦρ, fire, and ἀγρεύω, I take.

Sea. The metal works of Sarepta were constructed near the boundary of the country of Azara, and afterward acquired a great celebrity. It is amid this tribe of Azara, where everybody was a miner, according to Father Cobius,[1] that the mines of Carmel were situated.[2]

21. At the same epoch, Inachus was carrying from Egypt to Peloponnesus a colony which founded the kingdom of Argus. Three hundred years later, Cecrops, starting from the same point, founded the colony of Athens; while Cadmus, coming from Thebes in Egypt, constructed a city bearing the same name in Bœotia, and assumed the sovereignty.

Greece at this time begins to be peopled everywhere; and being the result of Egyptian emigration, receives the knowledge of the industrial arts, which the Egyptians, one of the oldest people in the world, had possessed to a great extent.

22. Cadmus gives the knowledge of bronze to the Greeks,[3] and discovers the mines of Mount Pangæus.[4] Minos introduces in Crete the art of working iron.[5] This metal is soon discovered at Mount Ida,[6] where, according to Grecian tradition, it was

[1] Aser Metallifossor (Præsid. Joh. Christ. Wickmanshausen, ling. or. prof., Wittemb., 1722).

[2] Κέρμηλος, ἀφ' οὗ χαλκὸς γένεται (Hesych.).

[3] Hygin. fab. 274.

[4] Plin. lib. vii. cap. lvii.

[5] Clem. Alex. Strom., lib. i.

[6] Marm. Oxen., ep. 11.

disclosed after a conflagration of forests.[1] The Dactyli, priests of Cybela, take hold of the discovery, and introduce the use of iron and steel in Phrygia. Prometheus, owing the knowledge of fire to a thunderbolt striking a wood, creates forges in Scythia, while Vulcan builds iron furnaces in the island of Lemnos.

23. The use of iron and steel had been known all over the East for a long time, and these metals had been employed for various purposes, when Lycurgides proscribed gold and silver, and made the coins of iron.[2]

24. The Phenicians, those skilful navigators of antiquity, who worked the mines of Eubœa, already exhausted at the time of Strabo, were the first to pass the Straits of the columns of Hercules (Straits of Gibraltar), where they founded the city of Gades,[3] in order to have a harbor in their travels to the Cassiterides. Soon after, they were followed by the Chalybes, a tribe of Armenia, celebrated in the manufacture of steel,[4] who gave that industry to

[1] Arist., περὶ θαυμασίων ἀκούσματον, 1157 E.; Diod. lib. i. 5; Strab. lib. i. 3; Athen. lib. i. 6.

[2] Several ancient peoples have used iron for the same purpose: the Clazomenians, according to Aristotle (lib. ii. des Econom.); the Britons, according to Cæsar (Comm., lib. v. 13); the Byzantians, according to Pollux.

[3] Now Cadiz.

[4] From them the Greeks gave to steel the name of χάλυψ.

Western Spain, while the Greeks were introducing it on the eastern shore and in Italy.

25. Diodorus of Sicily speaks of the island of Æthalia as being rich in iron ores, which the natives were working, "breaking them into pieces before melting, in order to extract the iron which is part of the stone." The stone being once broken, it was thrown into a furnace made for the purpose. A violent fire melted it, the parts became aggregated, and the product was a large metallic sponge. The ore thus transformed was sold to merchants, or rather exchanged for merchandise, and exported to Dicæarchea and other places. There these sponges were forged and transformed into various implements, which, afterwards, were peddled in the different parts of the known world.

26. The art of working iron had advanced so far as to alloy this metal with bronze for statuary. The celebrated artist Alcone had a Hercules made of hardened iron or steel. In Rome, wine cups made of steel were dedicated to Mars the Avenger, and deposited in his temples.

27. The Romans do not appear to have modified much the Grecian furnaces for iron and steel. During their stay in the Peninsula, they applied themselves especially to the working of precious

4

minerals, and we must say that they displayed a great skill in it.

28. On the other hand, the Moors were properly the manufacturers in Spain; they gave to the working of iron and steel such an importance, as to make it possible to foresee the preponderance which these metals, the most useful agents of civilization, were to attain ten centuries later. They overspread the Pyrenees with small hand forges, and made of these high mountains covered with forests, a centre of fabrication, the workmen of which were so celebrated that they supplied blacksmiths to all the adjoining countries.

29. During the first century of the Christian era, Calatayud, the ancient Bilbilis, near Moncayo,[1] was celebrated for its manufactures of steel.[2] Pliny says that the waters of the Salo, which ran around the city, were well suited for hardening metals.[3]

[1] The ancient city of Bilbilis was situated upon a mountain near the actual site of Calatayud. Upon the Monte Bambola, at about one mile and a half from Calatayud, the remains of that ancient city are yet to be seen.

[2] The great poet Martial, who was born at Bilbilis, about the year 40 A. D., says, in Ep. 55, lib. iv.:—

Nostræ nomina duriora terra
Grato non pudeat referre versu:
Sævo Bilbilim optimam metallo,
Quæ vincit Chalybosque, Nóricosque.

[3] Lib. xxxiv. cap. xiv.

30. While the working of iron and steel was increasing in the north of Spain and extended to the Aquitanian side of the Pyrenees, the Romans were introducing the metallurgic art into Germany, where it received many improvements.

31. The ancient metallurgists then worked two kinds of iron ore: a pure ore, which required no preliminary treatment; and an impure ore, which was broken, assorted, washed, and roasted.[1]

The pure ore was smelted in low furnaces, similar to a Catalan forge, and produced iron or steel almost by chance, when that ore was rich and without vein stone.

The impure or refractory ore, after it had been ascertained that it could not be worked in the same low furnaces, was smelted in much higher furnaces, square, and with an opening on the top into which the smelter threw the materials broken into pieces of the size of a nut. This is the origin of the *stück ofen* or *high bloomery furnaces*, which are the beginning of our present blast furnaces.

32. We could scarcely believe that, in this metallurgy of transition which lasted the first part of the middle ages, the limestone and silica fluxes mixed pell-mell with charcoal and ore, had not produced a raw metal in sufficient quantity to discover acciden-

[1] Agricola, *De re metallica*, lib. ix. p. 337 et seq.

tally the characteristic property of this new product. There is no doubt that we must go back to the twelfth century to find out the first idea of a blast furnace, and of the use of a raw metal (pig metal), the knowledge of which had been so long time delayed, only by the feeble power of the blast apparatus.

33. Among the Greeks and Romans, the bellows were made of leather;[1] they were moved by men or animals.[2] We find the same in ancient Spain.[3] Hydraulic wheels only came into use about the sixteenth century for pumping in mines, and as movers of stamping mills, and other apparatus appertaining to blast furnaces.

34. Agricola, who wrote at Schemnitz, A.D. 1546, fails to mention raw metal or blast furnace. We must hence infer, without fear of a mistake, that the art of casting raw metal into moulds is an invention relatively modern, which was unknown in Upper Germany.

35. Although this is not the place to inquire where the discovery of cast metal was first made, we cannot refrain from saying that all indications point to that invention having taken place near the Rhine.

[1] Quam folles taurini habent, quum liquescunt petræ ferrum ubi fit (Plaut., Ed. Schmied., v. 31, p. 885).

[2] Beitrage zur Geschichte der Erfindungen, 1, 3, p. 321 (Beckmann); Geschichte des Bergbaues der Alten, p 128 (Reytemeyer).

[3] De Hispaniæ antiquæ re metallica, p. 44 (Bethe).

36. In 1409, there was in the valley of Massevaux, between Riembach and Oberbruck, a blast furnace for smelting iron, which lasted only thirty years.[1]

37. That blast furnaces were extant in France in the middle of the fifteenth century is a well-known fact, which we will prove elsewhere by authentic data. We do not pretend to say that they were invented in our country, but we have certainly the right to take date, and to wait until we are shown records older than ours.[2]

38. The way the Wootz steel is made in Asia, the birthplace of the human race, brings to mind that the modern metallurgists, when they invented blast furnaces, have but imitated the processes of Persia, Salem, and Golconda. There cannot be more analogy than there is between these two methods, used in times and countries so far apart; one of these methods being older than the invasion of Alexander the Great, the other born scarcely four centuries since.

In India, from time immemorial, a magnetic ore,

[1] This valley is situated in the Departement du Haut-Rhin.

[2] Englishmen, who pretend to have invented the blast furnace, have no evidence going as far back to deduce. Mushet (Papers on Iron and Steel, p. 387) goes back only to 1540 A. D. to find indications of the manufacture of raw iron in the forest of Dean; O'Reilly (Annales des Arts et Manufactures, t. vi. p. 226) pretends that blast furnaces were extant in that country in 1450, but produces no evidence.

compound of oxide of iron and quartz, is smelted with charcoal in a furnace (*fluss ofen*) five feet high. This process was perpetuated in the Himalaya, and Mungo Park found it again at Kamalia, in Central Africa.

39. The oldest fact known in England in relation to the moulding of raw iron, is in the year 1547, when a Frenchman named Pierre Baud, established in that country, was smelting cast iron pieces for the English navy.[1] His workman and successor, Thomas Johnson, became celebrated for his skill and the perfection of his works.

40. As for the manufacture of steel, the eastern countries were far in advance of Europe, when the Greeks founded forges in the Peninsula, which afterwards were succeeded by the Catalan furnaces in the Pyrenees. In all these works, steel was extracted directly from the ore. Biscay, for a long time, enjoyed a great credit for its steel, and Bilbao had yet, in 1548, the privilege of supplying the English market with fine tools, such as engraving and point tools, punches, scissors, &c. The great market for these products had early awakened the cupidity of English manufacturers, who tried to counterfeit them. Bad faith went so far that, under the name of Bilbao iron, tools made of hardened iron without any value or quality were sold. In 1548, Parlia-

[1] Worthies of England, by Fuller, 1662.

ment was obliged to intervene and to prohibit this reprehensible fraud.

In the boundaries of Germany, and since the discovery of raw metal, people thought of refining pig metal, leaving in it enough carbon to make a kind of natural steel. The centre of this manufacture seems to be confined to the Alps.

41. Germany is also the first country where it was proposed to cement iron. Thence, this art came to France and was introduced at Newcastle on Tyne long before it was known at Sheffield, the present centre of that fabrication.

42. Sheffield cutlery itself does not date very far back; the first knives manufactured in England were made in 1563, by Thomas Mathews, of London. Previously, that country imported its manufactured steel ware from Flanders and the bordering countries.[1]

43. The working of cast-steel, entirely unknown in France a few years ago, was introduced into England only in 1770, by Mr. Huntsman, of Attercliffe. For a long time it was kept a secret.

44. Puddling steel is an entirely recent invention; in England this discovery is claimed by Mr. Ewald

[1] Oddy's European Commerce.

Riepe, who took a patent in 1850; but to Austria must we give the honor of the invention.

The first experiments were made in 1835 at Frantschach, in Carinthia; MM. Schlegel and Müller took a patent in 1836; thence the process went to Cibiswald in 1849, and to Neuberg, Styria, in 1851. Inasmuch as the steel obtained by that process did not fulfil all the expectations, it was discontinued at these places, either by want of a market, or by discouragement.

However, in 1846, M. Bischof had made experiments at the Hartz, in a gas furnace, and MM. Weyerhammer had followed in Bavaria and at Limburg on the Lenne. They seem to have been discouraged by some difficulties. These obstacles were overcome only in 1849, in Westphalia, where some iron masters, by much perseverance, succeeded in producing cast steel, in such quantity as to have it become a regular manufacture in 1850. MM. Lehrkind, Falkenroth & Co., of Haspe, were enabled to exhibit at London, in 1851, puddled steel of good quality and low price.

Thus Englishmen are not the first in using this process; they have been, this time again, skilful imitators.

45. In France, steel works have much progressed. In 1833, there were but 69 steel forges producing 2850 tons of forge steel; in 1852 the production went up to 3938 tons. 25 cementation or convert-

ing furnaces produced, in 1832, 2964 tons of cemented steel; in 1852 that quantity was increased up to 9808 tons. The manufacture of cast steel, which in 54 works was not over 324 tons, was 4352 tons in 1852.

PRELIMINARY OBSERVATIONS.

We will see that steel is a compound of iron and carbon, and is more or less acted upon, mechanically and chemically, by various bodies and elements which have a tendency to modify its properties. These reagents are heat, oxygen, sulphur, phosphorus, water, and lime. It will be useful, from the beginning, to examine briefly these substances, mostly with regard to their action upon the metal we are considering.

I.

Heat.

46. The cause of heat is unknown, but the phenomenon is felt by its effects.

Its first action upon steel is expansion. To understand this peculiar action, we must conceive that the molecules of steel are movable, and may be separated, leaving between them open spaces or pores, which are filled by molecules of heat, not perceptible as long as the temperature is low. This separation of the metallic molecules, kept apart by the presence of heat, causes the body submitted to that phenomenon to become extended in every direction, and thus increased in size. This is what is called *Dilatation* or *Expansion.*

If heat is given off, the dimensions of steel must, of course, diminish; the pores disappear, the molecules approach each other, and the solid becomes more compact, more dense and heavier in proportion to its volume. This phenomenon is called *Contraction.*

The expansion of steel (not hardened) is $\frac{1}{927}$, or 0.001079 of its volume at 100° centigrade.

47. The dilatation of steel increased to 1300° produces another phenomenon. The molecules, by being separated, cause in the whole mass a state of softness which in a short time will show another phenomenon called *Fusion.* At a temperature of 1400° centigrade, the least fusible steel will melt. Pig metal will liquefy between 1050° and 1250°, and iron at 1600°.

48. In proportion as heat penetrates the pores of steel and expands it, it produces on its surface different colors. On this point we will not speak now, but will return to it again (386).

49. The materials producing heat are called *Fuels.* Their calorific value is very variable—*i. e.*, they produce very different amounts of heat, and, on that

account, are worth studying. But this is not the place for minute accounts, which will be found in the *Manuel du Maître de Forges*, edition of 1858. It will be sufficient to know—and this is important to the manufacturer—how much heat each kind of fuel will produce. As a standard in the comparison of fuels, a calorific unit, called *calorie* (in French), *unit of heat*, or *kilogramme degree*, has been created. A kilogramme degree is the quantity of heat necessary to increase 1° centigrade the temperature of 1 kilogramme of water. By comparing the various kinds of fuel, according to the weight of each requisite to produce the same effect, it is easy to form a list, where each fuel is represented by a certain number of units of heat. But this way of reckoning the calorific power is entirely theoretical or absolute, and differs widely from the useful effect or calorific power obtained in practice. We must then look for the latter.

The total heat of saturated steam at 112°.5 centigrade is equal to 640.8 units of heat, or kilogramme degrees. The problem consists in finding out how many kilogrammes of water at 0° will be vaporized at 112°.5 under a pressure of 1½ atmosphere, by 1 kilogramme of fuel, and to multiply the weight by 640.8 units of heat. The following numbers have been found by this method:—

		Vaporized Water.	Fuel. Not Dried.	Fuel. Dried.
Wood		3.20 to 4.21	2351	2939
Charcoal (pine)		6.40 to 7.13	4345	4864
Peat		2.34 to 4.08	1986	3256
Carbonized peat		6.43 to 7.24	4281	4550
Lignite		2.03 to 4.02	1933	3554
Bituminous coal.	Close burning	5.63 to 6.85	4149	4380
	Dry burning	5.58 to 8.18	4495	4548
	Smith's coal	6.84 to 8.07	4926	5011
Coke.	Ordinary	6.59 to 7.50	4582	4902
	For metallurgy	7.45 to 7.70	4857	5156

II.

Oxygen.

50. Oxygen is that portion of atmospheric air which sustains life and combustion, and at the same time oxidizes metals. This gaseous substance has such an affinity for iron, that if this is in a state of minute division and great purity, as with the Chenot sponge, it will combine with and oxidize it entirely.

51. This sponge, of which further notice will be taken (314), is an agglomeration of molecules of perfectly pure iron, which are kept together by cohesion, but may be easily reduced into powder. In this case, the division of the metallic material is extreme; the smallest spark will fire it—that is to say, will begin the combustion. Once begun, and air furnishing its oxygen all the while, the combustion goes on, great heat is produced, and all the iron is burned.

If, then, the weights of the sponge before and after combustion have been noted, we find ourselves astonished by an increase of weight of about 22 per cent.

52. This is caused by the oxygen of the air, which has combined with the iron, increasing its weight; 100 parts of the burned mass have then the composition:—

Pure iron	77.78
Oxygen	22.22
	100.00

53. Oxygen has such an affinity for iron that this metal will absorb it continually, without the help of heat or fire. Thus, if the oxidized sponge, with its 22.22 per cent. of oxygen, is left to rest for some time, and if its composition is then analyzed, it will be found that it has absorbed a fresh supply of oxygen, and the result of the analysis will be—

Pure iron	70
Oxygen	30
	100

These two degrees of oxidation are the limits; the first is the oxide *minimum*, or first degree of oxidation; the second is the oxide *maximum*, or the highest degree of oxidation.

54. True it is, oxygen has a great affinity for iron, but it has also a greater one for carbon, which will be spoken of hereafter. It follows that, when iron is at the same time in company with carbon and oxygen, curious reactions will take place.

55. Without heat, iron alone will be acted upon by oxygen, this being a gaseous substance which can move and be carried towards iron, while carbon is in a solid state and remains inert.

By the intervention of heat, oxygen will combine with the molecules of carbon, thus producing carbonic acid, while the iron remains pure and unacted upon.

At a high temperature and when iron is about to liquefy, there is a double reaction: the carbon combines all at once with the iron, making a carbide or carburet, and with the oxygen, making carbonic acid and carbonic oxide; but in this case, it is necessary that the quantity of carbon should be in excess of what is needed to saturate all the oxygen.

56. This double reaction takes place during the working in the blast furnace. The iron ore, which is nothing else than an impure oxide of iron, is in the upper part of the furnace, at a height called the *reduction zone.* The carbon, on the contrary, is in the lower part, and is the basis of the inflamed fuel. At a certain time, when the blast apparatus forces air into the furnace, the oxygen becomes separated

and combines with the carbon. *Carbonic oxide*, which is a compound of one atom of oxygen and one atom of carbon, is formed, and goes through the *boshes* up to the upper parts of the stack.

It is unnatural for carbon to combine only with one atom of oxygen; on the contrary, it has a great tendency to saturate two atoms of that gas. If this does not occur in the boshes, it is for want of sufficient oxygen to produce *carbonic acid* (1 atom of carbon, 2 atoms of oxygen); carbon must then remain in the state of carbonic oxide all through the space filled with coal, until it comes to the reduction zone, where the layer of ore is found. There the ore has so much softened that the metal is ready to lose its oxygen. Carbonic oxide combines with it, becomes saturated, escapes in the state of carbonic acid, and iron is left pure.

Such are the reactions in a blast furnace. We will soon explain how that iron, when going downwards, will become steel, and afterwards raw metal.

57. The oxidation of bar steel requires a certain temperature above the freezing point. In the polar regions, steel does not become oxidized. It seems that its pores must be open to allow the introduction of oxygen.

III.

Sulphur.

58. Sulphur is an enemy of steel. Even in minute quantity, it makes it cold and hot short. Its presence is due mostly to the quality of the ore used for natural steel or pig metal, or to the fuel in contact with the ore.

59. Iron ores very often are contaminated with pyrites or sulphurets of iron, which often cannot be eliminated. A long exposure to air and natural percolation, a strong and protracted calcination, are the only economical and industrial means in the power of the ironmaster. In the first case, the sulphuret is transformed into a sulphate which water dissolves readily and rain carries away; in the second case, the fire volatilizes the sulphur, and expels it in the state of sulphurous acid.

60. As for the mineral fuels, such as pit coal and coke, which indeed are often full of sulphur, there is no practical corrective; the best is to reject them, and to use only the pure ones. The choice is easy, and for that it is only necessary to examine the color of their ashes. Brown, fallow, or even yellow ashes are indicative of a fuel rich in sulphur; a red color is the sign of a maximum of that metalloid; white ashes show that the fuel is free of this impurity.

The presence of sulphur in pit coal is due only to the iron pyrites it holds. Ferruginous coals will be then more apt to contain sulphur than those entirely carbonaceous. Brown, red, yellow colors are due to the presence of iron, and denote also that of sulphur.

61. Iron bars for cementation, which are piled up alongside the walls of warehouses, are exposed to the inclemency of the weather, and mostly of rain. If they contain a considerable quantity of sulphur, a double chemical reaction takes place. Under the influence of the dampness of air, iron becomes oxidized, the sulphur also, which, combining together, produce green vitriol, thus making an advantageous purification. It has been found by experience that 0.000084 of sulphur, regularly distributed in 1000 kilogrammes (1 ton) of iron, would produce 734 grammes (0.000734) of green vitriol.

This is an important fact from which it can be inferred: that a long exposure to the air of iron and other materials used for the manufacture of steel, such as calcined ore, pig metal, fine metal, puddled iron, &c., deprives them of their sulphur, by transforming it into protosulphate of iron easily washed off.

IV.

Phosphorus.

62. The action of phosphorus upon iron and steel has some analogy with that of sulphur: by it, iron

becomes hot short, and steel cannot weld, the fusibility being too great.

However, in certain cases, phosphorus will suspend the noxious effect of sulphur by neutralizing it. Carbon will do the same by decomposing its compound.

63. The great inflammability of phosphorus makes it of small account in the manufacture of steel, in which it is rarely to be found. In the working of the blast furnace, it evaporates, or is transformed into an evaporable substance as soon as it comes to the upper parts of the boshes. It melts at 35°.8 centigrade (Fahrenheit $63\frac{14}{100}$). When sulphur and phosphorus are found together in materials submitted to metallurgic treatment under a certain temperature, there is a production of a sulphide of phosphorus, very inflammable, which will escape, producing often an explosion and illumination.

V.

Water.

64. Water is combined with a certain kind of iron ore, called *hydrous oxide of iron*, in the proportion of 10 to 15 per cent. Otherwise, water will act in the metallurgy of iron and steel, only on account of the great quantity of oxygen it contains, thus being able to produce the phenomena of oxidation, deoxidation, and decarburization.

65. Water, indeed, holds nearly 89 per cent. of oxygen, which is pre-eminently the gas sustaining combustion. This is four times as much as is found in atmospheric air. It would seem, at first, that water should increase the temperature four times quicker than the blast does; but this does not take place, because in water, the oxygen is retained by the hydrogen with more strength than oxygen by nitrogen in the air.

The two elements of water are kept together by the force of affinity, by chemical combination, and cannot be separated unless by decomposition, *i. e.*, by a chemical reaction where certain conditions must be met. In air, the two elements are mechanically mixed, and may be disunited by simple separation.

66. When molecules of heat are mixed with molecules of water, or, to speak more plainly, when water is heated, it is transformed into steam; *id est*, it passes to an aëriform state in which the molecules of water are kept apart by the molecules of heat. Water is not decomposed thus; every molecule has kept the same composition it had previously in the liquid state, but the intimate union, the affinity of the elements is lessened. Hence, at a high temperature, steam will promote combustion.

Atmospheric air, or the blast, is very readily decomposed at an ordinary temperature, if it comes in contact with a body for which one of its two elements has a great affinity; for instance: in contact with iron, it will be transformed into oxide; in contact with carbon, it will make carbonic acid. The decomposition of air will also take place more readily when its molecules are distended by heat. This explains the advantages of the hot blast.

67. We have already seen that water facilitates the separation of the sulphur in iron (61); it will also separate silicon.

68. Silicon makes iron cold short. It cannot exist in pig iron or steel, in the state of oxide (silica), on account of the presence of carbon: it is there in the state of silicon, the pure metal. When steel or pig metal are on the point of losing their carbon, silicon has a tendency to become oxidized; but if it comes in contact with water, instead of being transformed into silica, it will be dissolved into the liquid, and thus will facilitate the purification of iron.[1] This effect is the more striking when water is in the state of steam. If steam is made to pass through a furnace, over a metal rich in silica, all of this which is in contact with steam will be dissolved, and a large portion of it will be carried away and deposited upon the sides of the furnace where the steam escapes.

[1] This is one opinion. Many persons think that silicon is more readily oxidized than carbon.—*Note of Translator.*

69. The water intimately mixed with the ores or fuels will cause their decrepitation; this effect is produced by small explosions which occur when the material is submitted to a sudden heat which instantaneously transforms the molecules of water into steam. The sudden expansion produces an effect similar to that of a bursting boiler, and the phenomenon is repeated with each molecule of water. Among fuels, anthracite will be noticeable by its decrepitation and its sparkling, when particles will be flying around, having the form of scales with an appearance of cleavage.

70. We have said (61) how useful was water in expelling sulphur from ores, metallic matters, and fuels. We will not repeat it. We will say only that, when this liquid is in a hygrometric state in ores and fuels, these will generally require a preliminary calcination.

VI.

Lime.

71. Calcium, whether in the state of oxide or carbonate, has a very great influence in the manufacture of pig metal and in the management of a blast furnace; it is employed for neutralizing the bad effects of silica, and for vitrifying the earthy materials which turn into slags or cinders; besides, it will extract sulphur. Lime might be advantageously

used in the manufacture of natural steel and even cast-steel, if this was suspected to hold sulphur.

72. Nevertheless, it is only in the state of calcium, that lime is found in steel, and this very seldom. Under such circumstances, this metal will harden steel, the same as silicon and aluminium do, without in the least impairing its quality.

The substances we have so rapidly examined are not those only which are found in contact with steel. Carbon, silicon, aluminium, magnesium, and manganese have yet more influence on its good or bad qualities, and in some cases are quite a component part of the carburet. Consequently, we have thought it would be better to speak of them in the theory of steel, and we have postponed their description from the first section of this work to the chapter on the theory of carburets (155).

VII.

Iron Ores.

73. The manufacturer of natural steel or cast-steel requires to know the iron ores employed by himself, or by the ironmasters who furnish him with pig-metal or iron. We will briefly describe them.

74. Iron ores are oxides of iron mixed with foreign earthy matters.

75. Although there are a great number of varieties of iron ores, although there are two degrees of oxidation, it is possible to classify these minerals in a small number of species, whether they are considered in regard to the quantity of oxygen, or are examined in regard to their places of extraction. They are:—

1st. Carbonate of iron (sparry, spathic iron).
2d. Oligist iron (specular, iron glance, red hæmatite).
3d. Magnetic iron (magnetite).
4th. Hydrous oxide of iron (limonite, brown hæmatite, bog ore).

Magnetic iron is alone attracted by the magnet; but the three other species will become so after calcination.

76. They may be distinguished from each other by being streaked with a steel point, or by being pulverized.

The streak is gray with carbonate of iron.
" " red " oligist iron.
" " black " magnetic iron.
" " yellow " hydrous oxide of iron.

77. § 1. A distinct characteristic of the *carbonate of iron* is that the metal is in the state of protoxide, ferrous oxide, or minimum degree of oxidation. Its streak is gray, and it becomes magnetic by calcina-

tion. Aside from these characteristics to be found in all the varieties, it ought to be divided into two sub-species, which are very different from each other in place of extraction and composition.

These two sub-species are lithoid iron and spathic iron.

78. *Lithoid iron*, often called clay iron stone, black band iron ore, is found in the coal measures, where it alternates with layers of coal, forming small layers of nodules.[1] In its composition silica is found in large quantity; this points to a limestone flux for fusing it.

79. *Spathic iron* occurs in rocks of transition. Magnesia and oxide of manganese enter largely into its composition; hence, irons made out of it are especially good for the manufacture of steel.

80. The analysis of the two sub-species of carbonate of iron gives on an average—

	Lithoid iron.	Spathic iron.
Volatile matters . . .	35	38
Protoxide of iron . .	45	52
Silica	10	0
Magnesia and manganese .	0	10
Other earthy materials, .	10	0
	100	100

[1] In the nodular state it is generally pure; the other kinds intimately mixed with a coaly clay often contain pyrites and phosphoric acid.—*Note of Translator.*

81. When carbonate of iron is calcined, or after a certain exposure to the air, its original color, gray or light brown, turns to red, brown-red or brown-black. The cause of this is, that the metal, which was oxidized at the minimum, absorbs a larger quantity of oxygen by the calcination or by exposure to the air, and thus passes to the state of peroxide of iron, or of oligist iron, with the proper color of the latter. Its streak changes also, and turns red instead of gray.

82. § 2. *Oligist iron* is really an oxide at the maximum degree of oxidation, a peroxide of iron, ferric oxide. That which has been produced by the transformation of which we have just spoken, is only one of the numerous varieties of that species, whose most distinct characteristic is to give a powder or a streak having a bright red color.

83. Oligist iron presents all sorts of colors, from dull red up to the brightness of polished steel; sometimes it is crystallized, when the surfaces are bright, iridescent and with metallic lustre; sometimes it is lamellar, the laminæ of which are polished, giving it the name of specular iron, on account of its property of reflecting the rays of light like a mirror. It has also a foliaceous structure whose small laminæ have a steel-gray, bright red, and sometimes sparkling color. In some places it is powder-like, granular, or compact.

84. This species is one of the richest in iron. It is often mixed with silicious earth, such as quartz. Oligist iron does not contain water, is not magnetic, but becomes so after calcination; it does not effervesce with acids. Analysis gives the following numbers, on an average:—

Peroxide of iron . . .	71	to	93
Silica and earthy materials .	29	"	7
	100		100

This indicates a percentage of pure iron equal to 50 to 65 per cent.

85. One of the varieties of this rich ore is called *red hæmatite;* as indicated by the name, it is blood-red. Its texture is fibrous, radiated, often similar to that of threads of raw silk, radiating from the centre of prominences towards the surface. The texture is sometimes so compact, that burnishing tools are made out of it for polishing gold or gilt objects.

86. This variety, after exposure to air and rain, becomes hydrated. Its brown color, after a long time, turns to a more or less bright red, loses its silky lustre, leaves its mark upon fingers, and is transformed into *sanguine* or *red chalk*, from which the red pencils of carpenters are manufactured.

87. Red hæmatite is most generally superior to

all other varieties of oligist iron for its percentage of iron. It contains, on an average, from 80 to 95 per cent. of peroxide of iron, which corresponds to 56 to 66 per cent. of metallic iron. It has also a great tendency to unite with manganese.

88. The great amount of metal and the small quantity of earthy materials in oligist iron, do not make it very convenient for working in blast furnaces, where it would produce but a white metal. It is a first rate ore for the manufacture of steel in low furnaces, such as a Catalan forge. This variety of hæmatite is sought mostly for that manufacture, very likely on account of the large quantity of manganese with which it is associated.

89. § 3. *Magnetic iron* or *oxidulated iron* is a compound oxide made of the two preceding oxides; it is sometimes so pure, that Brard brought from Sweden a sample having the following composition by analysis:—

(Ferrous oxide) protoxide of iron . .	69
(Ferric oxide) peroxide of iron . .	31
	100

which indicates 72.44 per cent. of metallic iron.

90. It is from this rich ore that Sweden extracts those celebrated irons which are all exported to England for the excellent steels of Sheffield. It is

probable that the oxide of manganese found in the ore helps to make these irons so advantageous for that manufacture; even magnesia, which is generally found in these ores, might have a similar effect. An analysis of Swedish pig metal, made by Berzelius, corroborates this assertion:—

91.	Iron	90.80
	Silicon	0.50
	Magnesium	0.20
	Manganese	4.57
	Carbon	3.93
		100.00

92. § 4. *Hydrous oxide of iron* is very rarely employed in some forges to produce pig metal or iron for steel making. However, there is in Styria an ore which contains as high as 16 per cent. of water; it is true that, at the same time, it is remarkable for its percentage of manganese and magnesia.

93. In the Pyrenees, there is worked for steel a hydrated oxide called *brown hæmatite*, with a stalactitic structure. With it occurs a blue black oxide of manganese, which sometimes is as compact as the oxide of iron itself, sometimes like a velvet-black efflorescence on the outer surface, or in the cavities.

94. Hydrous oxide of iron, outside of the varieties

we have just named, and of one or two more which are sought for the manufacture of steel, offers a number of varieties suitable only to produce pig metal in blast furnaces. In passing, we will cite the *brown oxide* with a large metallic percentage, and showing sometimes metallic iron by filing; the *reticulated spongy, foliaceous, hydrous oxides*, which often contain phosphoric acid; the *botryoidal brown hœmatite*, having much analogy with red hæmatite, except the color; the *red hydrated oxide*, with an earthy appearance and seldom compact; the *brown ochre, pisolithic iron, granular iron, bog iron ore*, &c. &c.

VIII.

Fuels.

95. Generally, the name of fuel is given to substances which, being combined with the oxygen of the air, produce the phenomena of combustion. In the arts, the fuels are the substances employed to produce heat.

96. The decomposition of their elements, their chemical and mechanical reactions, and generally, all their alterations or transformations taking place only at a higher or lesser temperature, all these causes give to fuels a great importance in metallurgy. Consequently they should be carefully studied by the ironmaster and the steel manufacturer. Their choice has a great influence in the quality and cost

of the products; and their calorific value, nearly always, will determine their more or less advantageous employment.

97. In fuels, the most abundant element and the one which plays the principal part, is carbon. It has a great affinity for iron, and, as we shall see ere long, is the basis of steel (142). It has the double effect of increasing the temperature of the furnaces, and of acting as a reagent in the treatment of iron and steel.

98. Among the vegetable fuels employed in the manufacture of steel, the principal are wood and charcoal.

Pit coal, coke, and anthracite are the mineral fuels.

99. Three constituent principles are found in every kind of fuel, no matter what class they belong to:—

1st. *Carbon*, which is the basis and principal element of heat. The calorific power of this element, when pure, is 7800 units of heat or kilogramme degrees.

2d. *Hydrogen*, the basis of water, and found in notable quantity in wood, even when dry. Its calorific power is equal to 22,115 units of heat.

3d. *Earthy materials* or principles of ashes. They are silica, alumina, lime, magnesia, oxides of iron and manganese, to which we must add soda and potassa in vegetable fuels.

We do not notice here a certain quantity of oxygen, in small proportion, which acts outside of the combustion proper, without helping it.

100. Carbon and hydrogen are then the two elements proper for combustion.

Carbon, the solid element, is the true principle of heat; hydrogen, the gaseous element, is the principle of inflammability. These two elements are in an inverse ratio to each other: the more a fuel is rich in carbon, the less it contains of hydrogen; the more continuous is its heat, the less inflammable it is. Wood is very inflammable: it holds six per cent. of hydrogen, but has little heating power; it contains but 50 per cent. of carbon, while anthracite has 91 per cent. The latter has the greatest heating power of all natural fuels; it holds but 3 per cent. of hydrogen, hence it is the least inflammable.

101. The ashes have a peculiar action in metallurgy: if they proceed from mineral fuels and come in contact with pig metal or steel during their fusion, they impart to these two metals the metallic principles of which themselves are the oxides, *i. e.*, silicon, aluminium, calcium, &c. During the refining of pig metal, these metals are oxidized again; silica, lime, and alumina unite with iron and make it cold short. Then, they are noxious.

102. If the ashes proceed from vegetable fuels, such as wood and charcoal, the alkalies they contain will vitrify the earthy principles. Silicates (slags, cinders, scoriæ) are formed thus, which can be extracted mechanically, leaving the iron very nearly pure. This explains the good quality of charcoal irons, and why steel is preferably worked with that fuel.

Having admitted these principles, let us now describe the various fuels.

103. § 1. *Wood* is the most natural of all fuels; it is very likely the origin of all the others. Bituminous plants have produced pit coal; the trunks and the branches of trees gave rise to lignite; the leaves are the origin of peat. Trunks and more or less carbonized impressions of plants will be found down in the deepest coal measures.

104. The combustible part of wood is called *lignine,* or *woody fibre.* Lignine, whether from the trunk or the branches, and no matter from what species it comes, contains, after being perfectly dried, nearly 50 per cent. of carbon and 6 per cent. of hydrogen. This is the theoretical percentage; but wood holds a great deal of water. When freshly cut, the quantity may vary between 0.2 and 0.5. After long exposure to the air, part of the water will evaporate, but a certain quantity will remain always, unless expelled by drying at a high tempera-

ture. The wood employed in metallurgy always contains from 20 to 25 per cent. of water.

105. The calorific power of dried wood is 3600 units of heat; that of common wood, holding 20 to 25 per cent. of water, is 2750 units of heat, on an average.

106. The woods employed in the arts are divided into two classes: *hard woods* and *soft woods*. Oak, beech, elm, &c., belong to the class of hard woods, contain in the same volume a greater number of fibres, and have a closer texture. Pine, fir, linden, poplar, &c., belong to the class of soft woods.

This division concurs sufficiently with the calorific value of these different species:—

107. A cord of wood (4 stères or 4 cubic metres), one year cut, produces the following amount of heat:—

		Units of heat.
Hard woods	Oak	6,846,000
	Ash	5,974,000
	Beech	5,603,000
	Elm	4,487,000
Soft woods	Birch	4,102,000
	Chestnut	4,035,000
	Pine	4,263,000
	Poplar	3,069,000

By burning the wood according to its density, we ought to come to the same results obtained and claimed by MM. Clement and Desormes, *i. e.*, that with equal weights, all kinds of woods have the same calorific value.

108. § 2. *Charcoal* is the product of the combustion of wood in places or vessels more or less closed, where the access of air is more or less avoided.

109. The object of this carbonization is to concentrate into a smaller volume the quantity of carbon disseminated in the woody fibre of the wood. Although beginning to be formed early, the quantity of lignine is small in young plants; with age the quantity increases; and in the various species of wood, when 20 to 25 years old, the proportion of lignine or woody fibre is from 90 to 95 per cent.

110. This is the limit of its increase; the tree may continue to grow in size, but the quantity of lignine remains in exact proportion to its weight, if not its volume. The limit of perfection is between 20 and 25 years: during that period of time, carbonization will be most advantageous. A lignine, over 25 years old, will not be more economical for charring; but when less than 20 years old, the woody fibre will produce a lesser quantity of charcoal.

111. Felling wood for charcoal is done during the spring, before the sap has begun to move. A few months later, this might injure the growth of the new shoots which issue from the stump. Wood, when felled during the fall, will not keep well, and is liable to be filled with holes. The wood, cut in the spring, can be carbonized before the summer is over.

112. The carbonization is done in *heaps*, or *meilers*, or in furnaces.

113. The meiler is built upon a dry and well compressed area. The logs of wood are so arranged around a central stake as to make a half sphere, all around which are left openings to regulate the fire. The meiler is covered with earth, charcoal dust, and turf sods, in order to prevent the action of the wind, and the fire is so regulated as to give the wood the time necessary to lose its water, gases, and be charred without the contact of the air.

A meiler is in itself a true furnace; but, as it is not entirely air-tight, much more lignine, and, of course, carbon are lost in the operation, than in a regular furnace.

114. In heaps, 100 parts of wood rarely produce over 17 to 18 per cent. of charcoal, while in air-tight furnaces, the product runs up to 25 per cent.

However, the yield should be 35 to 36 per cent. What is the cause of such a loss?

115. By a careful examination of what is going on during the carbonization in heaps, we see four distinct periods or stages during the charring process:—

1st. The mass becomes heated, dampness is expelled, and much steam escapes.

2d. Inflammation appears; the wood is very dry, contracts, becomes very dense, and is transformed into red charcoal.

3d. Combustion is beginning; the meiler sinks down; scarcely any steam is to be seen, and the red charcoal has turned black. The operation is complete.

4th. If the combustion is allowed to proceed, all the black charcoal will be burned out, and soon, nothing but ashes will be left.

Summing up, we have—

1st Period. The sweating; dampness is expelled; it is the dry stage, and nothing is consumed.

2d Period. The volatile matters escape. This is the Red charcoal stage. There is 36 per cent. of it in the meiler.

3d Period. Half of the red charcoal is burned. This is the Black charcoal stage. But 18 per cent. is left.

116. It is evident that, out of these three periods, the most advantageous is that where red charcoal is produced, when the weight of the product is double that in the third period.

The same might be said in regard to the carbonization in air-tight vessels, where about one third of the fuel is lost.

117. It has been ascertained by experience that a slow carbonization will give one-third more product than a rapid one. For instance, with ordinary meilers, oak wood gives only 17 per cent. of charcoal by a rapid operation, while the product goes up to 25 per cent. when the operation is somewhat slow.

118. As for the calorific value of fuels, we give a table of the quantity of heat produced by 1 hectolitre of the following fuels:—

	Units of heat.
Walnut charcoal	292,000
Oak "	255,000
Ash "	219,000
Beech "	176,000
Elm "	167,000
Birch "	153,000
Chestnut "	146,000
Yoke-elm "	176,000
Pine "	160,000
Poplar "	109,000

The absolute calorific power for 1 kilogramme of these charcoals would be 6.095 units of heat.

119. An analysis of charcoal has given—

Carbon	79
Potassa	a trace.
Volatile substances . .	14
Ashes	7
	100

120. The specific gravity of charcoal varies with the kinds of woods. It is—

With Walnut charcoal . .	0.166
Maple " . . .	0.114
Oak " . . .	0.106
Pine " . . .	0.075

The weight in kilogrammes of a cubic metre of charcoal varies with the nature of the soil of the countries where the wood has grown. We find—

	Cher.	Vosges.	Pyrenees.
	Kilo.	Kilo.	Kilo.
Oak and Beech charcoal .	245	228	230
Birch " .	225	228	230
Pine " .	205	228	170
Fir " .	205	135	170
Chestnut " .	205	135	140

121. The quantity of alkalies contained in charcoals may be rated at—

Oak charcoal . . .	0.80	per cent.
Beech " . . .	0.50	"
Elm " . . .	2.00	"
Aspen " . . .	0.60	"
Fir " . . .	0.20	"

122. § 3. *Pit coal,* or coal, is the most important of mineral fuels, on account of the enormous quantities found all over the globe, and on account of its inflammability and its heating power. Profusely scattered at all depths of the crust of the earth, coal will outlive our forests, and will furnish to the wants of society a fuel which our woods will soon be unable to give.

123. In regard to its industrial uses, pit coal is divided into three distinct classes—caking or bituminous coal, cherry or semi-bituminous coal, splint or close-burning coal.

Caking or bituminous coal (Houille grasse, Fr.) contains bitumen, and swells when heated. It gives a spongy residuum called coke, which approaches nearly to impure carbon.

Cherry or semi-bituminous coal (Houille maigre, Fr.) contains little bitumen, and does not swell when heated. It gives a long flame, and, on that account, is sometimes called flaming coal.

Splint or close-burning coal (Houille seche, Fr.) does not hold bitumen, does not swell and cake, and produces little flame.

124. It is probable that the early deposits of these fuels were only of the bituminous kind, but that the eruption of igneous rocks has caused a sort of distillation, depriving the coal of part or all of its bitumen, and producing thus the cherry and splint

coals. What corroborates this assertion is, that splint coals always occur in the vicinity of metamorphic rocks.

125. From what has been said about the elements of combustion (99), we may infer that bituminous coal is more inflammable than cherry or splint coals, and that the latter contain more carbon. This is shown by the following analyses:—

	Coals.		
	Bituminous.	Cherry.	Splint.
Carbon	76.25	84.10	91.30
Hydrogen	8.10	4.45	1.10
Volatile matters .	10.83	5.60	1.50
Ashes	4.82	5.85	6.10
	100.00	100.00	100.00

126. The calorific power of these three kinds of coal is expressed by the following numbers:—

	Units of heat.
Bituminous coal	7800
Cherry "	7200
Splint "	6600

127. In regard to the weight of coal, its specific gravity is 1.089; therefore the cubic metre weighs 1.089 kilogrammes in the coal bed; but, after it has been extracted and broken, the same measure will weigh only 800 to 880 kilogrammes, on account of the free spaces.

One hectolitre of coal will weigh, practically—

	Kilogrammes.
Creuzot coal	79 to 80[1]
Decize "	82 to 83
Saint Etienne coal . .	83 to 84
La Taupe " . .	85 to 86
Anzin " . .	85 to 86
Combelle " . .	86 to 87
La Barthe " . .	88 to 89

128. Most generally, coal will contain iron pyrites, which it is important to remove. The presence of pyrites is ascertained in the ashes by their more or less dark color (60). Sulphur would greatly impair the quality of iron and steel. Coal also holds a certain quantity of water, which it is good to notice in practice.

129. § 4. *Coke* is the product of the carbonization of pit coal.

130. All kinds of coal are not adapted to coke-making. To produce coke, some bitumen is necessary, in order to cake the molecules of carbon with the help of heat. With a higher temperature, the volatile parts of the bitumen disappear. Nevertheless, it is possible to make coke with any kind of coal,[2] provided some bitumen is added to them in the shape of pitch, or of very bituminous coal-dust.

[1] These numbers will apply nearly to the number of pounds in a bushel of coal.—*Trans.*

[2] Even with anthracite.—*Trans.*

131. The carbonization of coal is similar to that of wood; it can be made in heaps or in ovens. Sometimes it is effected between walls built around the heap of fine coal, forming thus a kind of half oven.

132. In heaps, 100 kilogrammes of coal produce from 40 to 45 per cent. of good coke; between walls the yield is from 45 to 50 per cent.; and in ovens from 60 to 65. But the coke made in heaps is more compact, brighter, and more sonorous than that which has been made between walls or in ovens.[1]

133. The volume of coke, compared with that of the coal from which it is derived, is 30 per cent. greater with bituminous coal. With cherry coal, the volume of coke produced is smaller than that of the coal itself; but in large operations, where various kinds of coal are mixed, it is calculated that the two volumes of coal and coke are equal.

134. The theoretical calorific value of coke is less than that of coal. The latter will give 7.500 units

[1] As much as 70 to 75 per cent. of coke, as good as that obtained in heaps, has been produced in ovens. Ovens will give, with the same quality and quantity of coal, a yield of 70 to 80 per cent. superior to that obtained in heaps. Certain appearances of coke are eagerly sought for in some places, while in others they are a cause of discredit. The best appearance of coke, *cœteris paribus*, is its behavior during metallurgic operations.—*Trans.*

of heat on an average, while the former will produce but 6000 units of heat. This is due to the presence of hydrogen in the raw fuel. One hectolitre of coke represents 230,000 units of heat; one hectolitre of coal 630,000. It would appear that coke and charcoal would give very similar results, one hectolitre of oak charcoal producing 255,000 units of heat. However, the value of coke is superior to that of charcoal from soft woods.

135. The coke from gas works is too brittle and too light to be used in metallurgy; it weighs only 300 to 350 kilogrammes the cubic metre, while the manufactured coke weighs from 400 to 450 kilogrammes.

136. The object in carbonizing coal is to expel the bitumen and the sulphur;[1] but, as by this operation coal loses half the heat which would be produced by its complete combustion, it is difficult to understand such a carbonization in an economical way, unless all the volatilized products are collected, instead of allowing them to be lost in the air, as is generally done.

137. Nature, by removing the bitumen from the coal measures nearest to the eruptive rocks, has pro-

[1] One of the principal objects is also to produce a fuel able to resist heavy pressures, which will not crack or split, and will allow a free circulation of the reducing and heating gases. —*Trans.*

duced a natural coke in the shape of splint or close burning coal. This latter differs from the coke only by its specific gravity caused by the enormous pressure of the rocks above it.

138. § 5. *Anthracite* is a peculiar fuel, very rich in carbon and without bitumen. There is too much tendency, among scientific persons, to confound splint coal with anthracite. It occurs in the oldest rocks, it presents no impressions of organic matters, and has for its principal characteristic—decrepitation when heated.[1]

139. It is a powerful fuel, difficult to kindle, burning with little flame, and thus similar to splint coal and coke. Workmen give it the name of incombustible coal, on account of its resistance to inflammation.

140. Anthracite contains no less than 90 per cent. of carbon; it gives an intense and constant heat which can be increased by a powerful blast. It can be burned simply with a good draft, and is adapted to domestic uses, as may be seen in the dwellings of Grenoble.[2]

[1] This is to a high degree a characteristic of the anthracite of Europe. It chokes a fire by the multitude of splints it produces when heated.

[2] In that part of France, anthracite is made to burn with a long flame under boilers, by closing the ash-pit almost entirely

141. This fuel is very advantageously used for metallurgy in Wales and in Pennsylvania. In France, where enormous deposits occur, it is overlooked and scarcely employed for burning plaster of Paris and lime.

This is the end of the *preliminary observations* we have thought useful to the manufacturer of steel. Those persons wishing to make a more thorough study of them should read the Manuel du Maître de Forges, published this year (1858), by my father, and which is complete on that subject.

(except when cleaning the grate) with an iron door, and introducing in that ash-pit a small steam jet. The steam passing through the incandescent fuel, is decomposed into its two elements, hydrogen and oxygen. The first burns readily, the second and the air of the draft make carbonic oxide and carbonic acid. A long flame is thus produced. This might be advantageous in some cases, but the economy is doubtful.—*Trans.*

PART FIRST.

STEEL AND ITS THEORY.

Steel.

142. STEEL is an alloy of the pure metal, *iron*, with the metalloid *carbon*. Therefore, iron and carbon united, whether as alloy, or by chemical combination, produce steel.

Before describing this alloy and its properties, it is well to consider separately the two elements which, when united, make steel. This study is the more useful, inasmuch as their behaviors are very different.

143. Iron is an elementary body, very useful in the arts on account of its ductility, malleability, and tenacity. Without being entirely infusible, it will resist a violent fire, and will soften only at a very high temperature. It may then be welded with itself, this being a characteristic which makes it so superior to other metals. Its specific gravity, 7.788, is greater than that of zinc and tin, but less than

that of the other industrial metals. The hardness of iron is small, when that metal is chemically pure, but will increase when it is associated with foreign bodies, such as carbon, silicon, and manganese, which themselves are elementary and distinct bodies.

144. The purest kind of carbon is the diamond, which has neither ductility, malleability, nor tenacity. It is exceedingly brittle, and its hardness is unequalled. It is infusible, not sonorous, and its properties, in a word, are entirely different from those of iron. Its specific gravity (3.50) is about half that of iron. Pure carbon is of exceedingly rare occurrence in nature, and on that account is much valued as a gem; on the contrary, it is very abundant in the impure state, as in wood, charcoal, mineral coal, coke, &c., where it is the basis of fuels, and in soot, leather, oils, &c.

145. The union of iron and carbon follows the general theory of alloys; we may compare it to the union of tallow with beeswax, when one of these substances is poured into the other which has been previously melted. But what is the nature of the alloy? Is it a chemical combination in definite proportions or a simple mixture? Is it a solution of one in the other?

In steel we find no characteristics of a chemical combination by atomic weights.

146. 1. There are only five carburets or carbides of iron which are chemically combined, in definite proportions and corresponding in simplicity of formulæ, with the natural combinations. They are the bicarburet of iron, the sesquicarburet, the protocarburet, the bi-ferriccarburet, and the quadri-ferriccarburet. Their composition is:—

	$Fe.C^2$	$Fe^2.C^3$	$Fe.C$	$Fe^2.C$	$Fe^4.C$
Iron	69.58	75.31	82.07	90.15	94.82
Carbon . . .	30.42	24.69	17.93	9.85	5.18
	100.	100.	100.	100.	100.

147. The small number of analyses of cast steel, which are recorded, give the following results:—

Cast-Steel.

Iron	99.442	99.435	99.445	99.360	99.360
Carbon . . .	0.333	0.330	0.340	0.325	0.335
Silicon . . .	0.225	0.235	0.215	0.315	0.305
	100.	100.	100.	100.	100.

Cemented Steel.

Iron	99.325	99.375	98.959	98.830	98.835
Carbon . . .	0.450	0.500	0.789	0.866	0.885
Silicon . . .	0.225	0.125	0.252	0.304	0.280
	100.	100.	100.	100.	100.

In these numbers, nothing indicates a chemical proportion, or an analogy with the formulæ of carburets of iron.

148. 2. Another characteristic of chemical combinations is, that they take place between bodies of opposite electricities, so that, at the moment of combination, the two kinds of electricity neutralize each other and disengage heat. When the union of carbon with iron takes place, not only is the temperature not increased, but, on the contrary, it is lowered.

149. In the alloy of these two elements, there is a great analogy with what is termed *solution* by chemists.

1. During solution, one of the bodies first becomes liquid; it is the solvent, and afterwards receives the other, which dissolves in it. When the union takes place, the temperature is lowered. This is seen in the formation of steel, where iron acts as the solvent into which carbon liquefies, absorbing a certain amount of latent heat.

150. 2. But carbon cannot be introduced into the mass of iron in an indefinite proportion. A certain quantity may be admitted by degrees, and this limit once attained, there cannot longer exist an intimate union. An excess will be admitted only in a state of mixture. This is what is called *saturation*. This phenomenon will appear in any solution, and is an

acknowledged fact for every alloy. This is well demonstrated by what happens during the solution of sugar in water; the liquid may receive successive quantities of sugar without its clearness being impaired; but if a great deal be added, the transparency of the solvent will be lessened, and the excess of sugar, being insoluble in the already saturated water, will remain suspended in the liquid during a while, and afterwards, when the mixture has rested, will be precipitated at the bottom of the vase by reason of its specific gravity.

151. The same phenomenon will take place in the formation of steel. The iron is melted first, dissolves the carbon, and becomes saturated up to a certain definite proportion. The carbon in excess, which is added to it afterwards, modifies the texture of the steel, and produces a new substance called raw metal or pig-iron. Steel is produced at a middle temperature, as will be seen hereafter; for raw metal, a higher temperature is required.

152. Thus, in the blast furnace, when the carbonic oxide (oxide of carbon) has deoxidized the iron ore and set free the pure metal, this metal, which falls to the lower part of the boshes through fuel at a high temperature, becomes first a saturated steel. It is only below the tuyeres that the molten metal absorbs a new and indefinite quantity of carbon and becomes raw metal (pig iron).

153. Steel thus produced has a constant composition as regards its constituent principles, iron and carbon. On the contrary, raw metal produced by an excess of carbon has a very variable composition. More or less fusibility in the ore, a good proportion of fluxes, more or less blast, the management of the furnace, &c., all will cause different qualities of pig-iron, from white crystallized metal up to the darkest gray pig-iron.

We have said that, in the blast furnace, steel, which is the first carburet obtained, has always a constant yield of carbon. This cannot be doubted,[1] because in the saturation of pure reduced iron the affinities react alone, and without any human hand to disturb the chemical apparatus; but this is different when the proportions of carbon and iron are left to the care of the workman.

154. If nails and charcoal are put into a crucible, and if they are strongly and sufficiently heated, steel will be produced; but that steel will be more or less hard or soft, according to the relative proportion of its elements. Will it be inferred that there are many kinds of steel? This would be wrong. The following, we think, will demonstrate sufficiently that alterations in the properties of pure steel are due to an excess of iron or of carbon outside of the definite

[1] It would be better to say: this is probable, until there are proofs at hand; but this is rather difficult to ascertain with certainty.—*Trans.*

proportions. In the first case, there is a beginning of ductility; in the second case, there is a beginning of brittleness. Sometimes the whole mass will be transformed into raw metal.

155. Carbon, the basis of industrial fuels, when alloyed with iron, renders steel fusible at a temperature not very elevated. How can a substance infusible itself, give to iron a property it does not possess? This has no more been explained than the vitrification of silica by alumina and lime. On the other hand the iron, which is essentially ductile, cedes part of its ductility to the carbon, and the alloy, in definite proportions, possesses that quality mostly when it is heated. It is a remarkable fact that, when hot, the malleability of the iron should prevail, while, when cold, the brittleness of the carbon will predominate.

156. In steel, sometimes part of the iron will be replaced by manganese, which is a metal not to be found in nature in a pure metallic state, but which has a strange analogy with iron.

157. Manganese forms, with carbon, some chemical carburets in the same number and with the same proportions as iron. Its equivalent weight is 27.6, while that of iron is 28. Manganese can be cemented the same as iron, and will produce a kind of steel or raw metal, whose residuum, by the action

of acids, is similar to that left by those two products of iron under a similar treatment.

The following analyses of the five carburets of manganese offer a perfect analogy with the corresponding compounds of iron, previously noted (146):—

	$Mn.C^2$	$Mn^2.C^3$	$Mn.C$	$Mn^2.C$	$Mn^4.C$
Manganese . .	69.38	75.11	81.91	90.05	94.77
Carbon . . .	30.62	24.89	18.09	9.95	5.23
	100.	100.	100.	100.	100.

158. The silver-like color of manganese and its feeble metallic lustre give a white appearance to the iron with which it may be combined; and to steel it imparts a fine bright grain, but also brittleness. The manufacturers of cast steel, by adding some oxide of manganese in their crucibles, aim at keeping the carbon in an accurate and proper proportion. Indeed, the oxygen of the manganese seizes the excess of carbon, and the metal reduced will increase the hardness and fineness of steel.

Mr. Robert Mushet, who took out numerous patents in 1856 and 1857 for the addition of manganese to steel, not only claims that the former metal improves the quality of the latter carburet, but also asserts that malleable iron, by an addition of manganese, may be transformed into steel, without adding any carbon, or any iron carburet. But no fact has yet been shown to prove this assertion in a practical way.[1]

159. Manganese acts advantageously on steels impregnated with silicon.

This latter metal, unlike manganese, gives to steel a darker color. It is not found in the metallic state, but, like iron, its oxygenated compound (silica) is widely scattered, and is the basis of a great many rocks.

160. The silica is reduced by the carbon of the fuel, and the silicon, set free, unites with the carburetted iron, if this metal is there in the metallic state. Therefore, steel may absorb more or less silicon, making with it an alloy which has not been

[1] It is likely that the so-called steel of Mr. Mushet, Jr., is nothing else but an alloy of iron and manganese, similar to those obtained by his father, Karsten, Berzelius, and others, which alloys have a certain hardness, and may be very well hardened. If we do not mistake, the action of the manganese is entirely different: from the oxide of manganese, the affinity of oxygen for the metal would counterbalance the affinity of carbon for iron in the pure steel. The result would be, that each time the oxide would be in contact with the carburet, no reaction would take place; but as soon as the quantity of carbon is in excess of the saturating proportion, part of it would be expelled in the state of carbonic oxide, or carbonic acid, while the chemical steel would remain pure, with a probable addition of a small quantity of metallic manganese.

On account of this hypothetical reaction, we have said that oxide of manganese will regulate the proportion of carbon.

studied sufficiently, but has the property of giving body to the carburet of iron.

161. Berzelius and Stromeyer, having exposed to the highest temperature of a blast furnace a mixture[1] of iron, silica, and charcoal, obtained the following alloy:—

Iron	90.70
Silicon	5.70
Carbon	3.60
	100.00

This is nothing else but a raw iron saturated with silicon, very apt to produce cold short iron.

162. In the analyses we have already given (147), it appears that most kinds of steel contain some silicon, which averages 0.259 in cast-steel and 0.237 in cemented steel.

163. Clouet, having succeeded in alloying iron and silicon in the following proportions:—

Iron	99.20
Silicon	0.80
	100.00

[1] The mixture was—

Iron	300	or	58.14
Silica	150	"	29.07
Charcoal	66	"	12.79
			100.00

and this alloy becoming somewhat hard by the process of hardening, it had been assimilated to steel. Indeed, it has been also inferred that silicon was as necessary as carbon in the production of steel. It would have been more prudent to say, that the metal thus obtained was a genuine (*sui generis*) alloy, having the characteristics of steel, but which, otherwise, could not take its place.

164. Magnesium is the elementary metal of magnesia, and appears to act nearly the same as silicon in steel, although it is to be found there in smaller quantity. The quality so much sought for in the manufacture of steel, of certain kinds of Swedish irons, may possibly be due to the presence of magnesia in spathic ores. Indeed, magnesium gives body and strength to steel.

165. The same effect will be produced by aluminium, the basis of alumina, which appears to give to steel a very great hardness and a tenacity which cannot be surpassed, if we believe the metallurgist who asserts that aluminium enters into the composition of the Indian steel called *Wootz*.[1] The alloy of steel and aluminium is white, very brittle, and has a very fine grain. Karsten, who has analyzed many kinds of iron and steel, found only traces of

[1] When speaking of the manufacture of Indian steel (290), we will demonstrate that aluminium has nothing to do in its composition, but that its quality is entirely due to silicon.

alumina. He has observed, also, that aluminium renders iron cold short; but as what may impair the structure of iron, is, on the contrary, often favorable to the quality of steel, the presence of aluminium in the latter metal will be rather advantageous by increasing its hardness.

166. Manganese, silicon, magnesium, aluminium,[1] produce intimate alloys with steel, but are not essential in its manufacture. They add certain qualities, such as hardness, ductility, body, &c.; and, at the same time, they regulate (mostly the oxide of manganese) the proportion of carbon, but they are not indispensable. An excellent steel can be made by the simple alloy of iron and carbon: Professor Schaffautl, of Munich, has analyzed the steel of one of the hardest razors of Rogers, Sheffield, and could not find a trace of the above-mentioned metals.

We will leave then these metals and their alloys, and will consider only the steel made of iron and carbon.

167. These two substances, as we have already said, form an intimate, definite, and unchangeable alloy, which is sometimes called chemical alloy on account of the constancy of its elements. This alloy, once made, will admit in its mass a greater

[1] The alloy of Tungsten with steel, tried for the first time in Austria, is exceedingly hard, and has been found to answer well for turning tools, chisels, &c.—*Trans.*

quantity of one of the two elements; and without any change in its chemical nature, it may be modified, present a different appearance, and possess new properties.[1]

168. For a good understanding of the reaction of iron with carbon, we will take a bar of pure iron which will be submitted gradually to the action of a combustible body.

169. 1. The bar of iron is strongly heated in a blacksmith's fire, and immediately after is dipped into cold charcoal dust. The iron will absorb in its pores, distended by the heat, some molecules of carbon separated from the impure charcoal, and which will penetrate the metal to a short distance from the surface. The iron becomes brittle, hard, resists the action of the file, and may be hardened. However, this is not steely iron.

170. 2. *Steely iron*[2] is the product of an incom-

[1] It seems that the ancients had ascertained the fact of a primitive nucleus of metal, in which were afterwards mixed particles of a less pure material. Very likely this is the meaning of *nucleus ferri* of Pliny, which Dr. Lardner translates by *well-purged iron*, and Mr. Vergnaud by *nodular ore*. These translations do not seem to me to correspond with the meaning of the Latin naturalist.

[2] Steely iron being intermediate between iron and steel, is advantageously employed when steel is to be welded with iron, as in some tools, pieces of machinery, etc., requiring only a

plete refining of the pig metal, by which all the combined carbon is not burned out, and remains intimately mixed with the whole mass, thus making a beginning of an alloy. In the first case, carbon has been mechanically introduced into the iron; in the second case, carbon was already there. These two kinds of iron having been hardened, it will be found that a file will "take" the steely iron more readily than the cemented one, because, in the former, the small quantity of carbon is equally distributed in the whole mass, while, in the latter, the file will immediately come in contact with all the carbon which is on or near the surface. The steely iron is a beginning of an alloy, and the other is the beginning of a cemented product.

171. 3. If the bar of iron, when cold, is put in the middle of a certain quantity of cold charcoal dust, in a vessel perfectly air-tight, and the whole submitted to a strong and continuous heat, the iron will open its pores, will soften, and the molecules of carbon will slowly penetrate it from the surface towards the centre. At the beginning, it will be easy to follow the motion of the carbon, because its small black molecules will be seen inlaid in the gray texture of the iron; but when the pores of the metal are kept open by these molecules, carbon will

facing of steel. Many hammers, planes, railroad tires, etc. are made this way, mostly on the continent of Europe. —*Trans.*

soon be scattered all over the mass, changing the texture of the iron, which has then a peculiar crystalline appearance.

The cementation is now complete. The metal thus obtained is not steel yet, although it has that name; but it contains the elements of steel in a heterogeneous and mechanically formed mass, which resists the action of the file, and may be hardened. This metal is termed *cemented steel.*

172. Its peculiar characteristic is to be generally covered with blisters, or cavities caused by the expansion, at a high temperature, of molecules of carbon combined with the oxygen from the iron or from the air in the box, thus making gaseous bubbles in the texture of the cemented bar. In England this metal is called *blistered steel*, and, in France, *acier poule*, which latter name is a corruption of *ampoule* (blister).

173. This union—we might say mechanical—of iron and carbon, which has not, and cannot possess any of the characteristics of the true alloy, which presents the defects of a heterogeneous and spongy mass without body, loses part of those defects after tilting or laminating. A perfect homogoncousness is not thus obtained, but the texture being closer, the carbon is in more intimate contact with the iron, and the metal becomes fibrous, tenacious, and may be hardened. However, some doubt will always

exist whether the carbon is in the proper proportion to make, with the iron, the definite alloy of a true steel.

174. 4. To transform this mechanical mixture into a definite alloy, and at the same time to produce a complete homogeneousness, the cemented bar is broken into small pieces, about the length of one decimetre, and a certain quantity of these pieces is put into a refractory crucible, kept for several hours in an intense fire. The cemented iron melts, and produces a homogeneous mass of pure steel, if the carbon and the iron are in proper proportions.

175. If it happens that the proportions are not satisfactory, the workman throws into the crucible some charcoal dust, or substances rich in carbon,[1] in order to perfect the quantity of carbon. But, lest it should be in excess, some oxide of manganese is added, which has the property of restoring the equilibrium.

176. This new product is *cast steel.* It does not always possess the requisite proportions of a definite alloy, although it is near enough to be taken as such. When it holds too much carbon, this is

[1] Nearly every manager of cast-steel works has a secret. Some add old leather to the metal; some use soot; while others prefer plumbago, etc. The only result of the addition of these substances is to furnish carbon.

ascertained by its hardness and brittleness after hardening. If, on the contrary, iron predominates, the hardened steel will be "mild" and softer.

These defects are corrected by a refining process, which will be spoken of hereafter.

177. Let us consider, now, what is going on during the formation of raw metal. In the lower part of the blast furnace the air introduced by the tuyere combines with an atom of carbon, and ascends through the boshes in the state of carbonic oxide (oxide of carbon). This, which must be transformed into carbonic acid before it escapes at the upper part of the stack, absorbs another atom of oxygen from the ore, which itself becomes reduced to the minimum of oxidation, and soon becomes pure metal. Then the molten iron falls through the boshes, surrounded by carbon. It dissolves part of the carbon, becomes saturated, and is thus transformed into cast-steel.

If the slag or cinder is not in sufficient quantity to envelop every drop of steel, in order to protect it against oxidation when it is going to fall on the hearth; the blast of the tuyere will partly decarburize the steel, and the molten mass will be a very white iron, a kind of half-fine metal. If, on the contrary, the temperature is very high, the carbon abundant, and the slag in sufficient quantity, instead of decarburization there will be a supersaturation of carbon, and a dissolution of this latter substance in the steel already saturated to the proper point; this forms gray pig iron.

178. The metallurgists who desire to manufacture steel with raw metal, understand at once what remains for them to do. If they employ the fine metal, they must add carbon; if they use gray metal, they must burn part of its carbon. In a word, they must bring the carburet of iron back to the definite proportions of the pure steel, which is one and unmingled.

179. In the Alps, where a natural steel is manufactured from raw iron, no other way is employed. The carburet of iron is partly refined, and the operation is ended when the saturation of steel is nearly complete. This is only an approximation, because the appreciation of that degree of refining being left to the practical judgment of the workman, very seldom is the true moment of saturation obtained.

180. In the Pyrenees, a kind of natural steel is obtained by the direct method. The ore is directly treated; the iron reduced into raw metal without the knowledge of the workman,[1] is afterwards refined and transformed into a state somewhat like

[1] Although it has been denied that in the Catalan forges, the ore was transformed into raw metal before being reduced into iron, it is nevertheless certain that such raw metal may be allowed to run through the *chio* (opening for the escape of the slags) of the hearth, before the slags are tapped off.

steel by a peculiar mode of working. Nevertheless, it often happens that the workman is deceived, changes the direction of the blast, and, instead of obtaining natural steel, produces only iron, to his great astonishment. It is easy to understand the cause of this.

181. To resume: if pure iron is cemented and cast; if white metal is carburized, or if gray metal is decarburized; if the iron ore is made to produce steel more or less directly; and if pure iron and carbon are thrown into a crucible to obtain cast steel, it is always the same theory, perfectly clear at this time, and which laughs at all the mysteries practised by charlatans, and believed by the ignorant.

182. All the difficulty is the proper dose of carbon; there is yet no industrial way to ascertain this with certitude, and without experimenting. The practised eye of the workman, his habit of always treating the same material, are the only guides at our disposal. The result is, that every steel-works produces a different quality, and that there is a multitude of varieties of cast steel.

183. After these observations purely theoretical and of a more or less speculative character, the following extracts from a work on steel by Réaumur[1]

[1] Réaumur (René, Ant. Ferchault de), was born at La Rochelle in 1683, and died in 1757. In his works will be found a complete description, drawings, and all, of the actual processes for cementing steel and making malleable iron.—*Trans.*

will be read with pleasure and surprise. It is remarkable to see how this skilful metallurgist has been able to discover and explain the theory of steel; and we do not know which is the more wonderful, the sagacity and correctness of mind of the celebrated savant, or the stupidity of the steel manufacturers who remained so long without understanding him.

184. "The workmen who manufacture the large sized files use only iron. Nevertheless, they make them as hard as steel files. The gunsmiths will give as much hardness to many parts of a gun, made entirely of iron, by case hardening. In this process, which will be fully explained hereafter, the pieces having received their proper shape from the hands of the workmen, are put into sheet iron boxes with a mixture of different drugs.[1] These boxes are smeared all over with some earth, and put into a furnace where they are submitted to a fire more or less protracted, according to the size of the pieces they contain. These pieces being taken off the fire, the workman dips them when red hot into cold water, where they harden the same as steel.

185. "Why does iron become able to be so hardened by such an operation? In seeking for the

[1] We translate as nearly as possible the old French of Réaumur.—*Trans.*

cause, I have ascertained that the first layers of metal are converted into steel. The iron files will act then the same as steel files; their teeth are of steel the same as the others. By experiments which it is useless to mention here, I have been convinced that this portion of iron is converted into steel, and of this workmen do not take notice; they use really steel tools, and believe them to be iron.

186. "The inferences I have drawn from this observation are: that the substances employed for case hardening could be made use of as a basis of proper mixtures for converting iron into steel; that, if those who case harden, submitted their pieces to a more protracted fire, they would become steel to the centre; this would be useless for the tools we have spoken of, which want hardness only in the first layers; but that observation was essential to me, who was endeavoring to transform iron bars entirely into steel.

187. "The bases of the compositions used for case hardening are powdered charcoal, ashes, soot seasoned with salts, and mixed with various substances of a vegetable, animal, or mineral nature. The secrets taught for converting iron into steel are founded generally upon these compositions; but each workman has his favorite ingredients, his particular doses of which he makes a mystery. After all, even had the workmen of Germany, England, and

other countries taught me their compositions, I would have nevertheless made the experiments which I will mention hereafter; I would not have been able to spare one. Independent of the advantage for the kingdom, the question in itself was important enough to be examined thoroughly. It was necessary to ascertain if the ingredients employed were the best, if in their stead the effect of some other would not be more sure and more rapid; to ascertain, for instance, if certain salts are worth the preference they enjoy; if some others, which might be used more successfully, have not been neglected; if in these compositions there were some substances which should be discarded as injurious or at least useless. It was necessary to be able to determine the exact proportion of every substance, to try if it might not be possible to make steel something better than is done to day; to see how far steel could be perfected. At last, it was necessary to systematize the mode of operation in converting iron into steel, to have this art known and easy to practise by workmen; but this art is to be found before it can be described. That object could not be attained without a number of experiments seemingly enormous; I have dared to undertake them, and I shall be well satisfied with the work they have entailed upon me, if they prove of some advantage to the public.

188. "We cannot afford not to begin by giving

an idea of the way the substances necessary for converting iron into steel are employed. Ordinarily they use boxes or large square crucibles, into which are inclosed the bars of iron which must be transformed into steel; some persons have these boxes or crucibles made of sheet iron, some use cast iron, while some others use only clay crucibles. Instead of boxes, a few have furnaces built for the purpose, where long bars can be placed. Whatever be the converting apparatus, these bars are cut into lengths proportionate to the size of the vessels; they are laid down, and a layer of the composition necessary for transforming them into steel is put between each layer of iron. The crucibles once filled are covered, luted, and submitted to a violent fire which is more or less protracted, according to the construction of the furnace, and according to the quantity and the thickness of the inclosed iron. The question was to make experiments which would show the effects produced upon iron by various substances separated or mixed together in different proportions, which would envelop that iron in a fire while it is heated at a temperature sufficiently high and protracted; with this in view, I had made a quantity of small clay crucibles square or oblong. All the crucibles for one operation were equal and similar; I inclosed in each crucible pieces of iron of the same quality, and equal in weight and in size; I gave them an equal heat, as near as could possibly be done; I surrounded the iron of each crucible with a different substance or with a mixture of various substances. Thus, the different changes in the iron were entirely due to the difference of substances, all other conditions having been the same. I have often used crucibles which held only one-half or one-quarter of a pound of iron with the surrounding stuff; thus I was enabled to despatch thirty or forty experiments at once in a rather small furnace. If I had begun by experiments on a large scale, the revenues of a powerful state would have scarcely paid for all the trials I wanted to make. Therefore I will say, in passing, that most of those who have tried to convert the irons of the kingdom into steel, have failed because they began the work on too grand a scale. We think that some of them had the basis of the secret; but before they could know what was to be added or suppressed, according to the nature of the irons they had to employ, or according to the construction of the furnaces they were obliged to use, they always wanted to begin by converting at once a large quantity of iron. Their first experiments were so expensive, that before they were able to finish all those necessary to correct the proportions of their compositions, they had expended all their own means and those of the persons with whom they were associated.

189. "I began by trying eight different mixtures. Case hardening gave me the idea of some of them; I added some which I found printed, and one which

Mr. d'Angervilliers, attentive to the welfare of the kingdom, had found in Germany, while commanding at Strasbourg.

190. "That first experiment was at least as successful as I expected; the irons from all my crucibles, after fifty-nine hours of fire, were more than half converted into steel; heated a second time, their transformation was complete. To be sure, they were not such steels as I wanted; some were coarse, some scarcely harder than iron, some others were fine grained and hard, but they could not resist the hammer, and it would have been impossible to work them. However, it was enough to show me that I was in the right direction, but that it was necessary to distinguish what was wanting in some of my compositions, and what was in excess in some others. Every one was to be analyzed, to know the effect of each constituent principle, which different substances were afterwards to be combined in different proportions. But, in order to forget nothing, and to go as far back as possible, I thought it was necessary, first, to ascertain if the iron which had been heated a long time and violently, without being exposed to the direct action of the flame, would not acquire thus the properties of steel, or if a continuous fire alone had not caused part of the changes I had observed. To be certain of the fact, I inclosed pieces of iron in different crucibles with inactive or nearly inactive substances. In some, the iron was enveloped with potter's clay

similar to that of the crucible, some with lime, some with plaster of Paris, some with burnt bones powdered; while with others different kinds of sand, leached ashes, and powdered glass were employed. All these experiments taught me that fire alone was not able to transform into steel the iron which was surrounded only by earthy materials nearly inactive. However, several of these substances had different effects upon the iron, which are worth being noted, and which might be useful.

191. "Lime, for instance, or calcined bones, far from imparting some of the quality of steel to iron, made it softer under the file and the hammer; and this observation might subsequently be applied to uses as important as the conversion of iron into steel.

192. "But a second observation, more peculiar on account of the preceding one, is that the plaster of Paris, which in itself is nothing else but the lime of a kind of soft stone, and from which results similar to those from ordinary lime might have been expected, has produced very different effects. Indeed, it did not change iron into steel; but who would have supposed that it was one of the most powerful fluxes for iron? When I had given to the crucible, filled with plaster of Paris, as much heat as to the other crucibles, I found the bars of iron reduced into a round and somewhat flattened mass which had

taken the shape of the bottom of the crucible. When the heat was not powerful enough to smelt the iron, this was entirely divided into scales which could be separated with the fingers, leaving only in the centre of the bars some fibres of soft iron. These scales could be broken the same as forge scales.[1] I have sometimes covered the crucibles where I had put plaster of Paris, and I saw a singular phenomenon: the plaster would boil and spirt out of the crucible, the same as a liquid, but a great deal higher. They were true boils, true jets of a fine powder, because the plaster had remained powdered such as when put in; the crucibles I had filled with plaster of Paris were nearly always broken before a great heat had been attained. After this experiment, I have tried if ordinary lime or calcined bones would not help the fusion of iron, and I could see nothing of the kind.

193. "The iron which had been surrounded with sand, such as that found at Fontenai-aux-Roses, and which is much esteemed by the founders of Paris, seemed milder after it had been taken off the crucible; it would acquire its previous hardness only after dipping in cold water. This experiment shows that blacksmiths may, without fear, throw this kind

[1] The reader will see by this paragraph, what was the chemistry of that epoch, and the difficulties Réaumur had to encounter. We do not suppose that, now, anybody would use plaster of Paris (sulphate of lime), even as a flux for iron.—*Trans.*

of sand upon the iron they do not want to burn in the forge fire, and that, without increasing its resistance to the file or the hammer.

194. "Although the iron, during this experiment, had not assumed any of the qualities of steel, we must note in this, and several other cases, that it had undergone some changes caused mostly, I believe, by the fire. The iron bars which were fibrous lost their fibre, and the bars whose structure was lamellar had their laminæ smaller.

195. "The pieces of iron which were enveloped with potter's clay, English and soft clay, remained also pure iron. However, they seemed to resist the file more than those put into other compositions.

196. "Leached ashes had the same effect as potter's clay.

197. "Glass approaches nearly to the nature of sands; it holds, indeed, salts which render it more fusible, but it keeps also the salts it has dissolved. The iron in some crucibles was covered with glass passed through a very fine sieve. This iron became a little harder, but without turning steel at all.

198. "The remarkable point in this experiment was, that the bars, which when put into the crucible were black, dirty, and somewhat rusty, came out

perfectly clear. The steel which most readily gets rid of its scale by hardening, is not so white in the place where it has been hardened the most. The glass had melted, had soaked, and we might say, had washed the pieces of iron; it had absorbed all the dirt without touching the scales, the volume of the iron not having decreased sensibly. Several arts require an iron perfectly clean; the scouring or pickling is made with sour liquids as in the case of the iron used for tin plates; it is possible that, instead of these liquids, a process similar to the preceding experiment could be made use of; should it succeed, such a long and tiresome work as in tinning iron or making tin plates, would be avoided.

199. "But in regard to our principal object, the result of the previous experiments is: that iron could not be converted into steel by heat alone; that heat is not helped by inactive substances of a too earthy nature, and devoid of oil or salts; and that the earths themselves, do not contain anything which might further the conversion of iron into steel.

200. "The persons who case harden, require the juice of several plants for hardening iron; many of them employ a great deal of garlic in their compositions. Never were the most savory sauces seasoned with as much garlic juice, as were the inert matters surrounding the iron of some of my cru-

cibles; but this seasoning was not active enough, and did not change the nature of the iron.

201. "I have tried afterwards what might be the action of seeds and fatty matters upon iron. I have saturated, with several kinds of grease, such as tallow, oils, mostly linseed oil, the earths and the lime which I had previously ascertained to be without effect. This pasty mixture was used for wrapping the iron of different crucibles. I found out by these experiments that oils alone cannot act upon iron to convert it into steel. It happens, indeed, that these oils are burned sooner than wanted; and although, to prevent that rapid burning, I had my crucible luted with great care, I could not perceive any change in the iron towards steel.

202. "I have also tried the effect of salts, whether by inclosing the iron in various kinds of salts, or by mixing a large quantity of these salts with earthy and inactive substances. These experiments have taught also that salts alone cannot give to iron any property of steel; all their action was to cut the fibres of soft iron, without giving it any power to become granular and hardened.

203. "But I have seen that this result, which could not be obtained by fire alone, neither by fatty or oily matters alone, nor by salts alone, might be attained by oils and salts mixed in certain proportions. It

is known that soap is precisely an oil thickened by alkaline salts, so as to become solid. I have mixed soap in different proportions with earthy substances; the iron inclosed in that mixture was half transformed into steel; *i. e.*, the lower part was so, while the upper part had remained iron. If the transformation was not complete, this was not due to the want of activity of the soap, but to its melting, thus acting only on that part of the iron with which it was in contact. The iron, whose nature had been changed, became indeed a very inferior steel; but, such as it was, there was proof that the conversion of iron into steel is to be expected with a mixture of salts and oily matters.

204. "After that, I went on trying substances naturally rich in oils, and in salts; I have tried first these substances without any mixture. I have put in some of my crucibles powdered charcoal; in some pit coal; in some soot, as it comes out of chimneys, or after letting it burn. In other crucibles I put horn, burned to the charring point, but not to ashes, which was powdered and afterwards sifted; and old leather burned and treated the same as horn. I have also tried the excrements of various animals, such as horses, chickens, pigeons, whether burnt or unburnt. I found that each of these substances had the power to change iron into steel, and this might be expected from the oils and salts they are impregnated with. But all of these substances are not

equally powerful. Charcoal, soot, old burned leather, may alone change iron into a fine and hard steel; but generally that steel is difficult to work, and after being forged remains full of cracks and flaws. However, these materials require a fire somewhat protracted; the action of soot and old leather is more rapid than that of charcoal. Horn, so much vaunted by steel-makers, did not seem to me to be more advantageous than soot; the effect was even less. Ashes do not make the iron difficult to work, but transform very little of it into steel; and that steel is so coarse that it is not worth the name. Pigeon dung produces a fine steel, but harsh; *i. e.*, when forged at a high temperature, it would break and fly off under the hammer.[1] Horse and chicken dung had scarcely any more effect than ordinary ashes. Pit-coal, previously powdered and sifted, had a very rapid effect; it had diminished considerably the volume of the iron, and had corroded and transformed it into hard and fine, but harsh steel.

205. "As a general result of these last experiments, it appeared to me that several of the ingredients above mentioned would enter into convenient compositions for converting iron into steel, and that some should be avoided, or their action somewhat moderated, such as those which produce a harsh

[1] This is due to the phosphorus found in certain quantity in the dung of birds.—*Trans.*

steel. On the contrary the action, too slow or too feeble, of some others was to be increased; and for that, it was necessary to try if the addition of some salt would not make them more powerful.

206. "Consequently I have searched how the action of these substances could be aided, and from what salts such aid might be expected. The more experiments are complicated, the more difficult it is to come to a decision on the causes of even their success. Thus, it was more difficult to decide upon the effect of every salt, than upon the other materials I had tried. The salts, as we have already seen, do not aid in the transformation of iron into steel, when they are alone or mixed with matters of a too earthy nature. By other experiments I learned what was the action of charcoal alone. I took it as a basis, and I have tried what would be its effect according to the kind of salt with which it was mixed. It is in that way I thought proper to try first the effect of the different kinds of salts. I took an equal weight of each of them, which I mixed with a much greater quantity of charcoal. All the weights were equal for every crucible, and of course the pieces of iron were equal. Afterwards, I made similar trials with the same salts, but giving them as a basis, and for a change, a mixture of soot, ashes, and charcoal, whose proportions will be indicated hereafter.

"The salts tried in these two different ways, gave

nearly the same results; and here is what I found the most striking in these experiments repeated several times:

207. "It seemed to me that powerful alkalies helped the conversion of iron into steel, but that they caused that steel to be difficult to work, to be full of flaws, and incapable of welding or drawing out. This I saw when using different kinds of soda, those of Carthagena, of Alicante, and potassa, &c. The natron of Egypt, which seems to possess the nature of an alkali, and which some chemists give as an example of an alkali not made by the hands of man, gave me also a harsh steel.

208. "Other salts would appear as hindering rather than helping the effect of charcoal; such was borax. I doubted also if alum or green vitriol had helped much in the transformation of iron; and I had some certitude only after having used them in a much larger proportion than the other salts.

209. "A peculiar effect of some salts is, that the steel they produce is not lasting. Such steel, which after being forged and hardened once, had a fine grain; when forged and hardened a second time, would scarcely have any grain. However, this singular effect was not constantly produced by the same salts. I mean that, when wishing to reproduce such steel with the same salts, I have not always

succeeded. The salts which gave me sometimes a steel so little lasting have not the same nature, and this makes the phenomenon more difficult to explain. These salts are sal ammoniac, sel de verre,[1] green vitriol, saltpetre concentrated by tartar,[2] or salt found after burning two parts of tartar mixed with one part of saltpetre. This latter salt has produced a steel difficult to work, the same as with all other alkaline salts.

"It has been objected that steels made from wrought iron would lose their fine grain the more they are worked; but this is not a general defect. It belongs to steels made with salts bearing analogy with those we have just spoken of. Steels from wrought iron will stand nearly the same as those made directly from the fusion of ores, when they have been treated with proper ingredients.

210. "The most important conclusion I could draw from my experiments with salts, was that among all others, common salt or sea salt is the most suitable to convert iron into a fine and hard steel, easily forged, and which does not deteriorate by working. Rocksalt, or the salt extracted from the boilers where saltpetre is refined, although having the same nature, never succeeded as well as the salt extracted from the water of the sea. Although I believe that the

[1] We do not know if "sel de verre" is soluble glass (silicate of potassa or soda) or sandiver (dross separated from glass).

[2] Black flux.—*Trans.*

salt extracted from the mother liquors of saltpetre could take the place of that extracted from sea water, I state with faithfulness what appeared to me, when I say that I succeeded better with common salt.

211. "In order to have more complete experiments with the salts, after having tried the effect of dry salts, I made some attempts with fluid salts, the spirits of salts. I have saturated with aqua fortis the charcoal with which I wanted to fill the crucible, until it had acquired the consistency of a soft paste. The iron covered with this paste has become a kind of steel, which remained such after the first hardening; but after a second forging and hardening, it became iron again. If we had not taken as a rule, in this first part of the work, to avoid every argument, this experiment would be a fit occasion to explain why steels made with certain salts will not last, as would be the case if they were produced with charcoal alone. I thought it was useless to go further with experiments on spirits of salts. It would not be convenient, in practice, to employ them; the expenses would be greatly increased. It is to be feared, also, that steel produced with those spirits, no matter what they are, would not stand the fire like that made with dry salts. Moreover, there would be a great evaporation of the spirits inclosed in the crucibles.[1]

[1] There are a great many other reasons for not using "such spirits," which the reader will understand without further explanation.—*Trans.*

212. "Besides the salts, I thought I should try if it might not be advantageous to employ various mineral substances, which are great fluxes for iron, and might, of course, be suspected of changing its texture. Some of these substances are pointed out as excellent for some kinds of hardening. Such are antimony, arsenic, ordinary sulphur, and verdigris. But no matter how the first three substances were employed, I found them good only for spoiling the iron or the steel. As for the verdigris, used in small quantity, the same as with the salts, it did not appear to me as producing such bad effects as might have been expected. It did not prevent the welding of steel, which is contrary to the prejudice of workmen, who think that everything holding copper will render iron impossible to be worked.

"The texture of the iron which had been surrounded with charcoal dust, mixed with antimony, had been changed; but it was not steel. The laminæ were no longer bright, neither were the fibres like those of iron, nor the grains like those of steel. The molecules had an intermediate appearance; they were flatter than the grains of steel and more raised than the iron laminæ; they were of a dull color, while the fibres of iron are bright.

"Ordinary sulphur, used in the same dose and with the same quantity of charcoal, as the preceding substances, has turned a soft iron into an intractable iron, and has prevented its conversion into steel by the charcoal. But, when I had mixed the same quantity of charcoal with a weight of the acid of sulphur proportionate to that of the sulphur in the preceding mixture, the iron was transformed into a coarse steel, difficult to weld.

"After having tried all the substances which I regarded as capable of acting upon the iron; after having ascertained those which should be entirely rejected, and those which could be employed with some success, it remained to try what might be the result of the active materials differently combined and in various proportions, and among these combinations, which was the most advantageous. With all these experiments, it was not easy for the most advantageous compositions for converting iron into steel, to escape my attention. Indeed, the number of combinations was large, but not so considerable as might appear. It is not necessary to go forward by insensible degrees, when sensible effects are to be produced; a physical precision lies between somewhat extended limits.

213. "After all these experiments, the compositions which appeared to me as answering the best, required only some powdered charcoal, ashes, soot, and common salt. But from these materials, mixed in different proportions, various compositions can be compounded. One, which I consider very proper for converting iron into a very fine and very hard steel, is made of 2 parts of soot, 1 part of charcoal, 1 part of ashes, and $\frac{3}{4}$ of one part, or something

less, of common salt, *i. e.*, if 16 pounds of soot are employed, 8 of charcoal, 8 of ashes, and 6 or only 5 pounds of common salt should be added.

214. "I prefer this composition, when it is necessary to convert into steel the irons having the best properties for this purpose. In another part of this work will be found the method of discerning the characteristics of these irons; but this same composition is not the best adapted for certain kinds of iron; by it, the steel would be too difficult to forge, to weld, or to draw, and after having been worked it would remain scabby. These kinds of iron require a less active composition; the following is used: 2 parts of ashes, 1 part of soot, 1 part of charcoal, and about $\frac{1}{4}$ of part of common salt.

"This latter composition might, as the former one, be successfully employed with irons most adapted for becoming steel; it will convert them, the same as the other, into good steel, but its action is slower. When this mixture is used, the operation is ended only after a great deal more protracted fire; this reason alone would make the first composition preferable, and we might add, also, on account of a superior degree of fineness given to the steel.

"Moreover, it will be seen, by what will follow, that this mixture may be always freely employed with various kinds of iron, although the steels thus produced are somewhat difficult to work. We will give the proper remedies for correcting its bad effects, and these remedies will not cost much more in time and in charcoal, than what is wanting in the composition. Generally, in prescriptions, no change whatever in the dose is allowed; this is nearly always the case with secret givers. We would imitate them, and this we do not wish, if we failed to give the information, that between the two compositions we have just given, there is an infinity of intermediate mixtures which may be successfully used. If we have determined with such precision the doses of the two preceding compositions, it is because the workman must have a basis to stand upon; because these doses show the limits between which it is proper to remain. Too far from them, there would be danger of producing coarse steel, or of making the operation too protracted. If, for instance, in the first composition, the dose of ashes was diminished or entirely omitted, it would be difficult to find irons which could, by it, be converted into steels easy to work. If, on the contrary, the quantity of ashes were increased too much, if it were to enter as three-fourths of the composition, and if the other fourth were to be divided between the charcoal and the soot, a much more protracted fire would be necessary to turn the iron into steel, or a much greater quantity of composition; and often only a coarse steel would be produced. But when proportions intermediate between the two stated limits are employed, no inconvenience will arise. For instance, one-third of soot, one third of ashes, and one-third of charcoal, with the quantity of common salt of one of the compositions, will produce a mixture which will not fail. But, if an iron having all the qualities for becoming a good steel is at hand, the first composition is better, for the reasons already explained; and, if the iron to be used has only some of the requisite qualities, it is more secure to use the second mixture. This is enough for a direction in practice; we will add only as a rule that, the more oily matters there are in the composition, the more danger there is of producing a steel full of flaws and difficult to forge; but the formation of steel is more rapid. The oily matters are found mostly in soot and charcoal; therefore, their quantity will be lessened by diminishing the dose of the two latter substances, or by increasing the proportion of ashes. This latter is principally employed for moderating the effect of the two other ingredients; it acts also by its alkaline salts, which are not in sufficient quantity to produce the bad effects we have pointed out, when speaking of the action of various salts.

"In order to obtain a greater certitude, relative to the bad effect of a too large proportion of oily matters, I soaked with linseed oil the substances of the first composition; the steel was rendered very difficult to forge, while, *cæteris paribus*, it would not have been so, if oil had not been added to the composition.

"Also, the dose of common salt we have indicated, is not so essential that it could not be changed; it could even be entirely omitted, but the operation would be slower; common salt makes it more rapid, and contributes to the hardness and fineness of steel. In the absence of common salt, a greater quantity of composition would be needed for the same amount of iron. The dose might be increased also; but up to a certain point it is injurious; if, for instance, twice the quantity is used, it is to be feared that the steel will be full of flaws, whether this effect is produced by salt itself, or because it helps the absorption by iron of the oily matters. However, an increase of common salt never seemed to me to produce such bad effects as an increase of oily substances.

"I have put into a crucible powdered charcoal alone, *i. e.*, without salt, or any other substance, but in large proportion, considering the weight of iron. This iron was converted into fine steel, but after a length of time nearly double what would have been necessary with the first composition, and this steel was full of flaws, after having been forged.

"When I introduced into my mixtures inactive substances, or nearly so, such as potter's clay, sand, or lime, I have stopped or hindered the effect of the active materials, according to the greater or smaller dose of the inactive ones. This was to be expected. However, if it was necessary to convert into steel some kinds of iron having too much tendency to become harsh steels, it would be possible to render them tractable, by regulating the effect of the active substances by some absorbing material. If, to our composition made of: 2 parts of ashes, 1 part of charcoal, 1 part soot, and three-fourths of a part of salt, we add one part of ordinary lime, or better still, one part of bone-lime,[1] *i. e.*, one part of bones burned and reduced to ashes, we have a composition by which certain kinds of iron have become steels easy to forge, while with any other composition they would not have been able to bear the hammer. The proportion of inactive matters may even be increased. I have sometimes converted iron into steel after having mixed two parts of bone-ashes with one part of wood-ashes, one of charcoal, one of soot, and the ordinary dose of common salt. But, after all, it is better not to try to change into steel irons which require such correctives in the mixture; if their proportion is too considerable, they entirely prevent the success of the operation. For instance, I have tried a process published in a book of *Secrets for the Arts*, printed at Paris, by Lambert, in 1716, t. i., page 12, and which did not succeed, on account, I believe, of a dose of quicklime too large in comparison with the other ingredients. This process requires one part of soot, three fourths of a part of oak-ashes, one-fourth of a part of eggs mixed; the whole is to be boiled in twelve parts of water, until these twelve parts are reduced to four. The pieces of iron are dipped in it, and afterwards made into layers alternating with other layers, composed of three

[1] Beware of the phosphorus in the bones.—*Trans.*

parts of charcoal, three of quicklime, one of soot, and one-fourth of one part of dried salt. This fine process left my iron very soft, which result I attribute to an excess of quicklime.

"Sometimes I have added one eighth of a part of lime to my ordinary compositions. In such small dose it never was injurious; it was even useful by diminishing certain kinds of blisters, of which we will speak hereafter, and which sometimes raise on the surface of the iron. A dose of plaster of Paris, smaller than that of lime, *i. e.*, one-twelfth of one part, is even more efficacious to prevent this phenomenon.

"Powdered glass, which some persons employ in their compositions, has scarcely any other use than to diminish these blisters; but it is no better than lime or plaster of Paris, and it would be a trouble in the manufacture to save enough glass for future use. Moreover, the defect which it remedies is so small that it should cause no uneasiness. The principal object in view, in large establishments, is to use only those materials which are easy to obtain.

"The above mentioned book, at page 81, teaches us another composition, one ingredient of which, for instance, would be difficult to obtain in large quantities. It is compounded of twelve parts of beech charcoal extinguished in urine, ten parts of horn, three parts of ashes of wood newly cut, and three parts of the powdered bark of pomegranate.

Where would manufacturers find their stock of this latter powder, which, moreover, I consider as more injurious than useful?

215. "But, coming back to the two compositions which we regarded as preferable, they offer the advantage of requiring only drugs which are easy to find everywhere, and which, excepting common salt,[1] are cheap everywhere; their preparation also is not expensive. The soot may be passed through a coarse sieve, but if it is fine, it is better. It is not at all necessary to have it calcined; this I found out, after using it burnt and unburnt. As regards the ashes, notwithstanding all that has been said upon the proper choice, they are always good, when coming from freshly-cut wood, whatever be the species of the tree. Ashes are sifted through a sieve not particularly fine; a similar sieve may be used for charcoal, after this has been powdered with a pestle. Any kind of charcoal may be employed, although the most active is that of oak. Charcoal from white wood did not seem to me to produce sensibly different results. Beech charcoal, which is intermediate between those from oak and white wood, may possibly be preferred; but if we speak openly, these differences are so difficult to ascertain by the most exact experiments, that such slight differences are of no importance in practice."

[1] In Réaumur's time common salt was verily heavily taxed.—*Trans.*

ANALYSIS OF STEEL.

QUANTITATIVE CHEMISTRY.

216. STEEL, we have said, is principally a compound of iron and carbon, and these two elements are to be found in it, in two peculiar states: either forming an alloy in definite proportions, nearly a chemical combination; or dissolved in this first alloy, only as a mixture.

217. If the proportions of iron and carbon in the definite alloy were exactly known, the analysis of the compound metal, in regard to these two elements, would be limited to the quantitative determination of one of them, the other being ascertained by induction and calculation. But, although it is rational to think that carbon is not over a certain quantity, this quantity has not yet been determined; or rather, there is in steel, as in pig metal, something which it has not yet been possible to decide upon, and which produces differences in analysis and throws a kind of mystery on that important matter. Until better informed, we shall be obliged to separate

in the assays, the *combined carbon* from the uncombined carbon or *graphitic carbon.*

218. Besides carbon, as has been stated already, steel contains silicon, magnesium, aluminium, manganese, sulphur, and phosphorus.

219. The steel which is to be analyzed must be previously reduced to a very fine powder, in order to facilitate the action of the reagents; on this subject, it might be useful to consult the new edition (1858) of the "*Maître de Forges*," where all the preliminary manipulations for these kinds of analyses are described accurately.

On account of the great hardness of steel, a file should not be used for obtaining the fine powder to be analyzed; fragments of the file itself would contaminate the assay sample. It is necessary to break the pieces of steel to be analyzed, under a cylindrical pestle, strongly hardened, fitting into a mortar of hardened steel similar to those used for pulverizing hard minerals. The result is passed through a fine sieve, and the coarser fragments are, if wanted, pulverized again.

220. This powder, having been made as fine as possible, is separated into three portions: The first is for determining the proportion of carbon; the second, for ascertaining the quantity of sulphur and phosphorus; and the third one will be used for find-

ing out the various constituent principles which, excepting combined carbon, may be left in the residuum.

221. 1. *Estimation of the Carbon.*—Some white sand is ground with some oxide of copper, and the mixture is calcined, in order to destroy all organic matters. 30 to 40 grammes of this powder are mixed with the same quantity of powdered steel, and triturated some time in an agate mortar. During this trituration, the mortar is put upon a piece of glazed paper, in order to lose none of the powder, which is afterwards mixed with six or eight times its weight of fused chromate of lead; the whole is then introduced, with the ordinary precautions, into a combustion tube, at the end of which are a few grammes of perfectly dry chlorate of potassa. The combustion is then conducted in the usual way; the carbonic acid passes through a chloride of calcium tube, and is absorbed and weighed in a Liebig apparatus holding a solution of caustic potassa, whose specific gravity is 1.28.

222. Although nitrogen has never been found in steel,[1] a search might nevertheless be made for it. Its percentage may be ascertained by mixing some

[1] Many metals, in the molten state, absorb large quantities of gases, such as oxygen, nitrogen, hydrogen, and keep part of them in cooling. See experiments by MM. Caron, St. Claire Deville, Frémy, and by Professors Abel, and Graham.—*Trans.*

of the assay sample with soda lime, and absorbing the products of the combustion by dilute hydrochloric acid, put in a Will and Warrentrapp apparatus. If there is nitrogen, chloride of ammonium will be produced, which can be estimated in the usual way, and which will give the quantity of nitrogen in steel.

223. 2. *Estimation of Sulphur and Phosphorus.*—The pulverized steel of the second portion is treated with fuming nitric acid at a moderate heat; it is soon attacked, and nitrous vapors begin to escape, without trace of sulphuretted hydrogen. The solution is then evaporated to dryness, and the residuum is treated again with very dilute hydrochloric acid. A small quantity of this solution is filtered and a few drops of chloride of barium added to it; after standing a few hours, if a precipitate appears, the remainder of the solution is filtered and treated again with chloride of barium. After entire settling of the sulphate of baryta, it is collected upon a filter, washed, dried, and calcined, thus giving the necessary elements by which the quantity of sulphur in the metal may be calculated and known.

The excess of the baryta in the solution is afterwards precipitated by a sufficient quantity of sulphuric acid, and separated by filtration. A solution of tartrate of ammonia is then added, in sufficient quantity to prevent the precipitation of iron by ammonia, and a great excess is required at this period of the operation. Sulphuretted hydrogen is also allowed to pass through the solution for several hours.

The liquid is then left to stand in a warm place, until it has acquired a light yellow color. It is now filtered rapidly, and the precipitate washed with water containing some sulphide of ammonium. After drying, the ammonia salts are expelled by calcination, and the residue is a compound of phosphoric acid with some lime, alumina, and alkalies,[1] which are mixed with a certain quantity of a mixture of carbonates of soda and potassa, and melted in a platinum crucible. The melted mass is afterwards dissolved in hydrochloric acid, and the phosphoric acid is estimated, by the usual way, in the state of double phosphate of ammonia and magnesia.

Instead of the method we have indicated for the estimation of sulphur, its proportion can be ascertained by dissolving slowly the pulverized steel in dilute muriatic acid and passing hydrogen through the solution. The gases are received in an acid solution of acetate of lead. If sulphur is present in steel, it will by this treatment be converted into hydrosulphuric acid, which combining with the lead of the acetate, produces a sulphide from which the weight of the sulphur is deduced.

[1] And more or less iron.—*Trans.*

This analysis must be conducted with extreme slowness; not less than eight to ten days are required for dissolving the proper quantity of steel. Pig iron requires ten to fifteen days, and wrought iron, four.

224. 3. *Estimation of Graphitic Carbon, Silica, Lime, &c.*—Let us take 30 to 40 grammes of the third portion of pulverized steel, which are treated with diluted hydrochloric acid in a glass balloon. By a gentle heat the iron is dissolved in a few hours, leaving dark or black flakes floating in the liquid. These are collected upon a filter previously weighed, washed, and dried at 100° centigrade. The increase of weight is noted, and represents the graphitic carbon with some silicate of iron and lime. The silica, iron, &c., of this mixture are estimated by fusing the whole with nitrate of potassa, mixed with twice its weight of pure carbonate of soda. The quantity of silica, lime, &c., being ascertained, by the usual method, the loss of weight indicates the graphitic carbon. In order to corroborate this result, another portion of steel is dissolved in dilute hydrochloric acid, and the flaky residuum is collected by filtration through the fibres of asbestos put in the narrow part of a funnel. After drying, flakes and abestos together are mixed with chromate of lead and oxide of copper, and treated in the usual way employed for destroying organic matters.

With proper care in the experiment, the results obtained by the two methods must be exactly the same. The graphitic carbon thus estimated (and which we will call *b* for greater convenience), being deducted from the whole quantity of carbon found in the first operation by combustion, we will have by difference the quantity of combined carbon *a*.

All the liquids are now evaporated to dryness, and treated again with diluted hydrochloric acid, which leaves behind undissolved a very small quantity of silica. This silica is collected and added to the silica previously obtained from the black deposit.

A small quantity of the solution is then treated with sulphuretted hydrogen; and if a dark precipitate is the result, the whole is to be acted upon by this gas.

If a dark precipitate is formed, it must be separated by filtration, and the metals it may contain are to be determined by the usual methods.

Generally, however, the small precipitate by the action of the sulphuretted hydrogen is some sulphur having a milky white appearance. This is collected upon a filter, and some nitric acid is added to the solution, which is made to boil till all the iron is peroxidized.

Afterwards, with ammonia added in small quantities, most of the iron is precipitated in the state of peroxide (ferric oxide). The last portions of this metal are separated by the neutral benzoate of ammonia;[1] and from the weight of peroxide obtained, the weight of metal in the steel is deduced.

225. After the weight of the oxide of iron has been taken, a portion of it might be used for discovering traces of chromium and alumina. For this, it is sufficient to dissolve this portion of oxide in hydrochloric acid, and to precipitate by an excess of caustic potassa which will dissolve these foreign substances, which are generally in very small quantities. If an excess of ammonia has not been added to the solution previous to the use of benzoate of ammonia, the precipitated iron will not contain a trace of manganese.

226. Before separating this latter metal, the solution and washings must be evaporated to dryness, and the ammonia salts expelled by calcination. After this treatment, the residuum has always a brown color, due to the presence of the oxide of manganese. It is dissolved in a few drops of hydrochloric acid; some ammonia is added first, and afterwards some sulphide of ammonium.

The precipitate of sulphide of manganese, after standing some time, and being slightly heated, is

[1] Succinate of ammonia, perfectly neutral, is generally employed, the precipitate obtained being less bulky.—*Trans.*

collected upon a filter. It may be transformed into sulphate of manganese, or dissolved again in hydrochloric acid, precipitated in the form of carbonate, and after calcination is weighed in the state of red oxide.

227. By ebullition the sulphide of ammonium is expelled from the solution from which manganese has been separated, and then an addition of oxalate of ammonia will precipitate the lime in the form of oxalate. This salt is calcined, and the carbonate of lime produced indicates the quantity of lime or of calcium in the analyzed steel.

228. If it is suspected that a steel contains magnesium, the presence of this metal may be ascertained by adding a few drops of phosphate of soda to the liquors filtered from the oxalate of lime. But it has never yet been found in steel, in quantity large enough to be weighed. Only traces may be found in iron. The presence of magnesia is ascertained only by qualitative analysis, and it may be entirely neglected in the quantitative analysis.

As for the alkalies, they are found in the solution from which lime has been separated; it is only necessary to evaporate this solution to dryness, to calcine and to weigh them in the state of chlorides. If, then, they are dissolved in a small quantity of water, and a few drops of bichloride of platinum are added, the potassa may be separated and weighed.

The chloride of sodium is found by a differential calculation.[1]

[1] The analytical chemist may modify several of the methods above indicated, in regard to reagents, weight and size of assay sample, rapidity of operation, &c.

New methods will be found in the Chemical News (American reprint, October and November, 1867, and April and May, 1868). But we do not yet know of any method for estimating the combined carbon, superior to that of combustion as in organic analyses.

At all events, a correct analysis of pig iron, steel or wrought iron, requires time and experience. Such an analysis, without accuracy, is not of much value.—*Trans.*

PART SECOND.

METALLURGY OF STEEL.

---

229. It results from the principles we have developed in the first part of this work, that the manufacture of steel consists in alloying iron and carbon in a definite proportion, thus producing a perfectly homogeneous alloy, whose properties are a great tenacity when heated, and an extreme hardness when hardened.

To produce this alloy, the metallurgist has at his disposal:—

1. Iron ore;
2. Pig-metal or raw iron;
3. Wrought iron.

230. With these materials, and charcoal which furnishes carbon, he can manufacture:

1. *Natural steel*, by deoxidizing the iron ore, and carburizing the reduced iron;
2. *Steel from raw-metal*, by removing the graphitic carbon (151).

3. *Cemented steel*, by introducing carbon into wrought iron (169–171).

As these three kinds of carburetted iron are not pure steel, and are most of them wanting in homogeneousness, they are submitted to a last operation, producing thus:—

4. *Cast-steel*, or the result of the fusion in close vessels of any one of the three kinds of steel we have just mentioned (174–176).

231. We shall next describe three different modes of fabrication.

We shall add to them the process employed in India, to produce a kind of cast-steel known under the name of *Wootz;* and we shall then examine several new processes of manufacture which attract, with more or less reason, the utmost attention from theoretical and practical metallurgists, such as the processes of Chénot, Bessemer, Taylor, and Uchatius. We shall end this second part by describing the working of Damascus steel and waved steel.

I.

Natural Steel.

232. Natural steel, or that obtained from the iron ore, is made in low furnaces, having an analogy with certain refinery fires called *Renardieres*, but known in the Pyrenees under the name of *Catalan fires* or *forges*. The shape of these furnaces varies with

every country, as well as their mode of working, thus making as many methods, with different names, but whose principle is the same. Hence, it will be sufficient to describe here the manufacture in a Catalan forge, to understand the methods in use in Navarre, Biscay, Corsica, etc.

233. The Catalan fire is prismatic. Its size depends on the quantity of blast it is possible to get; it increases with the power of the blowing machine. Too much blast in a furnace would be injurious to the uniformity of the mixture of ore with charcoal; while a small blast, with a large hearth, would produce a slow and laborious fusion. The quality of charcoal has also some influence on the dimensions of the hearth; light charcoal from soft wood requires, for burning, a larger hearth than would be the case with charcoal from hard wood.

Fig. 1.

234. However, in the same country, there is a great difference in the dimensions: at Gingla, the sides of the hearth are 0.43 by 0.59 metre, and the depth is 0.81 metre; at Sahorre, these dimensions are 0.70 by 0.70 metre for the sides, and 0.87 metre for the depth.

235. These fires are used for the production of iron or steel, not according to the wants, but rather per chance, which, with rare exceptions, is more the ruling power than science or experience in these iron works of the Pyrenees.

236. When working for iron, the tuyere must have a sharp inclination, in order to carry the blast directly into the fire, and to distribute it immediately upon the ore. When working for steel, the tuyere must be nearly horizontal, and in order that the ore which is smelting under the charcoal should not be decarburized too soon, its direction should be towards the cinders or slags, which must be produced in larger quantity for steel than for iron.

237. The projection of the tuyere over the hearth must vary with the size of the hearth, and the power of the fuel used. For a small fire and light charcoal, this projection is, necessarily, smaller than for a large fire and powerful fuel. It varies between 0.14 and 0.16 metre.

238. The angles of the hearth are rounded, thus producing a more or less elliptic shape. Consequently, the tuyere is somewhat turned towards the (*Rustine*) back, and thus, the blast is distributed with more uniformity than if the tuyere was towards the (*Chio*) front. Nevertheless, this disposition is variable with the state of preservation of the tuyere; because, if it is new, the blast is projected more directly and in one mass; but when it is worn out, the blast is scattered.

239. The ores worked for steel in the Catalan forges are spathic irons, rich in manganese. These ores are generally found all over the French and Spanish Pyrenees. They are roasted, and broken into lumps of middle size. The small pieces and the coarse powder are sifted and separated, thus forming what is called *greillade*.

240. For charging the furnace, charcoal is put on the left side near the tuyere, upon some charcoal already burning. The bottom of the hearth has been covered with a *brasque* made of charcoal dust well beaten down. On the right side, opposite the tuyere, and called *contrevent* (against the wind), a kind of wall or talus is made with the ore sloping towards the front (*laiterol*), where the tap hole (*chio*) is situated. Thus, the right side is covered with ore, and the left side or tuyere side is charged with charcoal. The mass of ore is also covered with

charcoal, on top of which a mixture of charcoal dust, coarsely powdered ore (*greillade*), and wet slag is beaten down. All these materials form a kind of arch, which is always kept covered with moist powdered ore, to prevent the flame escaping by any opening.

241. When the charge is complete, a blast is given gently and cautiously first, which is very slowly increased. It is a full hour after the operation has begun, before its entire force is allowed.

In proportion as the charcoal is consumed before the tuyere, the workman in charge replaces it by fresh charcoal, in order to prop the ore and to prevent it from tumbling down into the fire. By this mode of working, the ore has time to become deoxidized and ready for fusion.

When working for iron, after one and a half or two hours, the slag or cinders are run out from the tap hole. Raw iron will often run with them, and a great deal more of it would escape, if the ore which begins to agglomerate were not pushed towards the tuyere, in order to hasten its deoxidization. When working for steel, the cinders are not tapped so soon; the carbide which is forming must have the time necessary for a complete reaction, by which it becomes the definite alloy under the protection of slags, which prevent its entire decarburization.

The workmen have a habit of fattening the fire

(*engraisser le feu*), *i. e.*, of covering the mass of ore and charcoal with powdered ore. This method is very proper when ductile iron is sought for; but its defect is to thicken the slags and lessen their fluidity.

At a certain moment of the operation, the workman called *escola*, gathers the lumps of ore which are not entirely smelted, and helps their fusion with a rich slag, which must be saved for that purpose. Afterwards, he unites all the agglomerated pieces into one mass, which is, of course, very rough. This operation of forming the lump is called *baléger le massé*, in the Pyrenees. The asperities are removed with a crowbar (*palenque*), and the blast, from the horizontal direction it had during the forming of the lump, is brought to its former position. The fire is then much increased, in order to complete the fusion.

242. This being done, the blast is stopped, the charcoal is removed, and the lump of iron (*massé*) is uncovered. Immediately, all the forgemen come to help; one of them passes a bar through the tap hole under the lump; another, standing on the front side of the hearth, helps to raise the metallic mass with a hooked bar; while a third one grasps it between large tongs. When out of the furnace, the lump of iron is carried over the floor of the forge to the hammer (*mail*), where it receives a prismatic form. It is afterwards divided into two parts (*massoques*), one of which remains on the floor of

the forge, and is covered with burning coal, to prevent its rapid cooling and superficial decarburizing, while the other is shingled.

The next work is the cleaning of the hearth, in order to start a new operation, during the beginning of which the cold piece of iron called massoque is re-heated near the tuyere. This massoque is afterwards divided into two parts (*massoquettes*), which are also shingled.

243. Every smelting operation lasts from five to six hours, and the lump of iron obtained weighs 70 to 150 kilogrammes; generally, the charge of the furnace is from 210 to 450 kilog. of ore, of which 150 to 300 kilog. are broken lumps, and 60 to 150 kilog. are coarse powder. The consumption of charcoal is nearly equal in weight to the quantity of ore; *i. e.*, that for producing 100 kilog. of iron in lump (*massé*), an average of 300 kilog. of ore, lumps and coarse powder, and 300 kilog. of charcoal are required. It follows that the ore, which is generally very rich, and contains 55 to 60 per cent. of pure iron, produces no more than 33 per cent.

II.

Steel from Raw Iron or Raw Steel.

244. The pig metal used in the Department of Isère, and in the Alps, for the manufacture of *steel from raw iron*, sometimes called *raw steel*, is obtained

from spathic ores with charcoal. It is white; its grains are crystallized, quite large, and divergent, like those of fine metal. However, it is not much decarburized, because a drop of nitric acid put on it will produce a dark spot.

245. The fires employed for converting pig iron into steel, are a great deal larger than those in use for refining; they require only 6800 cubic decimetres of air, while over 10,000 are required for refining. The cause of such a difference is, that in the working for iron, about one-third of the blast is employed for decarburizing the pig metal entirely; while in

Fig. 2.

the conversion of pig metal into steel, only a small quantity of carbon is to be consumed. This explains also why the tuyere is horizontal, instead of dipping the same as in an iron finery.

246. The size of a raw-steel finery fire in the Isère, is one metre square and 1.50 metre deep. A sandstone is used for the bottom, and the sides are built with fire-bricks.

247. This mode of working requires four men; one head forgeman and three assistants.

248. The hearth is filled with fine charcoal, which is beaten down for two or three hours. This is called *making the brasque.* In the middle of this carbonaceous mass a hole is dug, 0.38 to 0.40 metre in diameter, and 0.50 metre in depth. This having been done, the hearth is filled with burning coals, covered with breeze (fine charcoal), and the blast begins to play. This preliminary heating is made use of for reheating there some blooms of steel, and drawing them afterwards.

The cinders are then taken out with a shovel, the cavity is cleaned, and again filled with charcoal. On top of this charcoal, 600 to 700 kilogrammes of broken pig iron are placed, care being taken that they should be arranged in battlement, and supported by a crowbar. The hearth is then inclosed in a wall of fine charcoal, previously dampened, and the whole is covered with charcoal and cinders. The fusion must last four hours.

During these four hours, the head forgeman has

nothing to do but to change the crowbars supporting the pig iron on top of the hearth, to probe the molten mass, and to add a little cinder and charcoal.

The pig iron remains thus in fusion during eight or nine hours, protected against the blast by a molten mass of slags, having a thickness of 0.15 to 0.16 metre. The melting and molecular arrangement of the elements is made quietly under the influence of a great heat. The head founder watches the operation attentively; increases the fluidity of the slags, when necessary, with some powdered quartz; prevents the thickening of the metal, by increasing the blast; or diminishes it, to produce thickening when the proper time has come.

The viscosity of the mass indicates that it has become steel. This is the moment for presenting the lumps to the direct blast of the tuyere, in order to burn the excess of dissolved carbon. When this refining has been rapidly done, an assistant takes one ball with the iron tongs, while another workman squeezes it with a sledge-hammer. This ball, or bloom, is then carried to the tilt-hammer, where it is shingled and forged into a prismatic shape, of which all the faces are drawn flat.

249. Twenty to twenty-one such tilted blooms (*massiaux*) are thus made.

The consumption for 1000 kilog. of pig metal is 3500 to 4000 kilog. of charcoal. The products are—

650 kilog. of steel,
150 " of iron.

In some localities natural steel is manufactured by another method.

250. The pig iron is refined in a special fire, where it is partially decarburized; after that, it is taken out in pieces about 0.03 to 0.04 metre thick. This done, a brasque is put into the hearth, which has the ordinary dimensions of a finery fire—about 0.55 to 0.60 metre. The tap-hole remains.

25 to 30 kilog. of pig iron are put upon the hearth, and the fusion quickly occurs. This operation requires about one hour and a quarter; during this time, the blooms of the previous operation are reheated, forged into bars, and immediately hardened. When the pig iron is smelted, it is left to stand, in order to allow the change of texture of the molecular elements to take place. The refining is conducted in the way we have indicated, and the blooms are taken out of the fire, and drawn the same as before.

In this operation, only a forger and his assistant are required, and they will manufacture, in twelve hours, 150 to 175 kilog. of steel. For 1000 kilog. of steel, the consumption is 1600 kilog. of pig metal, and nearly 4 *bannes*[1] of charcoal.

251. In Westphalia and in Silesia, the manufacture of steel differs from that in France; but the same

[1] A *banne* is a load of charcoal variable with the locality and the nature of charcoal. We do not know what is its value in weight or measure.—*Translator*.

principles will apply to all the methods for producing natural steel. A rapid fusion and a slow refining, such are the directions followed everywhere, and the theory accords with experience.

In these countries pig iron is not previously refined. Gray metal is often employed; but then the blast must be rapid and directed downwards, whilst white metal requires a horizontal blast. With gray metal, the first thing to be done is to destroy the graphitic or uncombined carbon; afterwards, more homogeneity is to be given to the remaining carbide. This is done by constantly stirring the molten metal, and thus preserving its fluidity; by this working, the carbon is thoroughly distributed in the liquid mass which has been kept covered with slags. For a thorough distribution of carbon, the metals rich in manganese are very advantageous. Therefore they are in great demand, and, as they are generally perfectly homogeneous, they produce the best raw steel.

In Westphalia and in Silesia, small plates of pig metal are put into the hearth, and are smelted with a small quantity of rich slags which are the first to fall, and thus cover the bottom which is made of a refractory sandstone. Other pieces of pig metal are put upon the recess plate, in order to profit by the heat produced; thence they are, one after the other, put vertically into the fire near the right side (the side facing the tuyere). They are replaced by the blooms of the previous operation, which compress the fine charcoal and prevent its dispersion. From this place, the blooms are brought nearer the fire, underneath the tuyere, where they are sufficiently heated to be drawn into bars.

Soon, the first piece of metal is seen to soften, sink down by degrees, and be liquefied. Its fusion is hastened, if desired, by bringing it nearer the tuyere, which may be inclined, if necessary. The motion of the blowing machine is increased, in order to produce a rapid blast, until perfect liquidity is produced. At this point, the blast must be decreased; some hammer scales are thrown upon the fire, and the mass is stirred until it again becomes pasty.

Afterwards, a second piece of *floss*, already red hot, is put vertically into the fire, and the blast is once more increased. This second piece, which weighs generally 15 kilog. (the weight of the first was only 12 kilog.) will, by smelting, render the pasty mass liquid again. If it appears that the mass has retained too much of the nature of raw metal, a small quantity of rich slag is added; however, this is to be avoided as much as possible. The blast is diminished as soon as the metal is liquid, which is stirred until a pasty consistence is obtained; but it is to be feared that the metallic paste will become too hard by refining, and will stick to the bottom of the hearth.

The third piece of plate, weighing 20 to 25 kilog., is to be treated the same as the preceding ones. The whole mass recovers its liquidity; some rich oxides are thrown into it while it is being vigorously stirred, and the action of the blowing machine is slightly slackened. If it is then seen that the metal sticks to the bottom, that it is becoming malleable, and that it produces too fluid slags, a very rapid blast is given, and the mass is stirred without interruption, in order to produce a brisk ebullition. When the stirring has been going on for some time, the mass falls down, and the metal is separated in the form of a cake; its working is ended only when it is no longer possible to drive a crowbar into it.

The fourth piece of pig iron, weighing about 15 kilog., is then put in the fire towards the centre of the metallic cake. This latter, being corroded by the raw metal only at its centre, is bored throughout, while the edges remain unacted upon. The blast, which is very rapid during the fusion, is to be moderated afterwards. The stirring is renewed, and continues until the new boil has ceased, and the mass has fallen. The fifth piece of raw iron is treated in the same way; often a sixth one is liquefied. During the last stirring, the blast must always be the strongest; however, the rapidity of the blast should be reduced if a hole is seen in the centre of the lump.

In order to prevent the formation of a layer of iron upon the lump of steel, the blast must be stopped at the proper time. This moment is ascertained, either by the consistency of the mass, or by the formation of slags sticking to the crowbar.

As soon as the blowing machine is stopped, the charcoal is removed, the lump is uncovered and left to cool awhile, in order that none of its fragments may be detached. Afterwards, with a sledge hammer, a crowbar passing through the tap-hole is forced into the hearth, and helps to raise the lump which sticks strongly to the sides. This lump is cut into six, seven, or eight blooms, having a pyramidal shape, the apex of which is towards the centre, because the steel is always a little harder towards the extremities.

The blooms of the previous operation are drawn out during the fusion. After being forged into square bars with sides equal to 32 millimetres, they are delivered to the refiners. But, as these bars must be reduced in thickness, it would be better to forge them into flat bars. This would be economy, and the steel would be improved.

252. The consumption of charcoal is very great; sometimes it runs up to 2.64 cubic metres for 100 kilog. of steel. The waste varies according to the quality of the raw metal, and the skill of the workmen. It will often be thought a satisfactory result when three parts of pig iron produce two parts of steel. If the raw metal is better, seven parts will give five of steel; and sometimes, when its quality is very superior, four parts of it will be sufficient for three parts of raw steel. Therefore, 1000 kilog. of

steel are produced by 1300 to 1500 kilog. of pig metal and 200 hectolitres of charcoal.

Every fire is attended to by a furnace man, a head forger and an assistant, because the working is not regular.

In some steel works, old iron is added to the molten mass, after the fusion of the fourth piece. The quantity of old iron is about one third of the weight of the bloom.

253. In the principality of Siegen, in Styria, in Carinthia, in Carniole, and in the Tyrol, different processes are in use. The steel works of Siegen refine the pig iron just as it comes from the blast furnace.

In Styria, the gray metal is converted first into iron, by the method of double fusion in use in that country. With white metal, the access of air is prevented as soon as the metal becomes lumpy. The Styrian steel, known under the name of *scythe steel*, is sometimes refined and forged into small bars. It is then called *mock*. The *mock* is milder than the *stuck stahl* or *German steel*. This is made in ordinary brasqued fires; the pig iron is put upon the brasque strongly beaten down, and is kept, the same as the blooms, with iron tongs, on the side opposite the tuyere. With a feeble blast the metal is smelted slowly, and some rich cinder is added to it. The fusion is slow in the Styrian method, and rapid in the German process. Therefore, in the one case, pig iron is refined by standing, and in the other, by working.

The steels manufactured for certain kinds of arms, swords, &c., and known under the name of *fine Brézian*, *ordinary Brézian*, *Roman steel*, require a great deal more care. The raw metal is first liquefied, and refined by stirring; but the lump is cut into several blooms which are reheated in a peculiar furnace and drawn afterward. The steel for arms is drawn into bars 0.025 to 0.030 metre thick; and the brézian into bars 0.010 to 0.015 metre thick.

In the principality of Siegen, the name of *edelstahl* is given to a hard and brittle steel manufactured at Mussen, by a process similar to that used in the north of Germany; the only difference is that white metal from spathic ores is employed, which, of course, shortens the length of the operation. Besides, hammer scales are added, and the cinders are tapped only at the moment of boiling. The *mittel kaehr* is an *edelstahl*, but not so brittle.

III.

Puddled Steel.

254. The puddling furnace for steel is somewhat different from the furnace for iron; the bed is not so large, but deeper; the sides are generally built with hollow cast plates, through which cold air or water is made to circulate, in order to prevent the rapid deterioration of these plates.

Fig. 3.

255. The raw iron employed is gray or white; that produced by spathic ores with charcoal being preferred.

256. The charge introduced into the furnace is from 140 to 150 kilog. of pig metal. A strong and rapid heat is produced, in order to prevent the decarburization of the surface; but as soon as fusion begins, the fire is lowered, and the heat is regulated with the damper.

This is the time to add a slag or cinder, which must cover the molten metal, which will remain liquid at a moderate heat, and at the same time, during the stirring, will somewhat decarburize the pig metal by burning the excess of carbon. The hammer or mill scales, and the cinders from reheating furnaces are convenient for this purpose.

When the raw metal is entirely smelted, the puddling process proper begins, and a flux made of peroxide of manganese, common salt, and dry clay is added.[1] This flux is mixed with the metal, and the whole is thoroughly stirred, mixed, and puddled in every direction. After a few minutes of working, the damper is raised, and a new charge of 20 kilog. of pig metal is put into the furnace, upon a bed of cinders, and near the fire bridge. This pig metal is then allowed to liquefy.

The mass remaining in the bed quickly boils, and the decarburization is heralded by jets of small, blue flame. The new metal, which had remained near the fire bridge, is then mixed with the boiling mass, which soon swells and raises. Small metallic grains are seen finding their way through the cinders, and indicate that the time for the puddling proper has come.

The damper is closed about three-fourths, and care is taken that during the whole operation the temperature should not rise above cherry red, or, at most, above the welding heat necessary for tilted steel.

The mass is then moved and stirred backwards and forwards under the covering layer of cinders. These signs of decarburization, *i. e.*, the small jets of blue flame of carbonic oxide, disappear; the small grains grow in number, become soft, and form a viscous mass, the temperature of which is cherry red.

[1] Some other ingredients have been proposed, such as chloride of calcium, nitrate of soda, &c.—*Trans.*

This is the time for increasing the fire, in order to maintain a constant heat during the operation. Next, the damper is completely closed; the metallic mass is entirely covered with the cinders, and undergoes the reaction necessary to the formation of steel, which reaction is facilitated by the presence of manganese. At times, the same as when puddling iron, a ball is formed underneath the cinders, and is taken out to the hammer. Bars, plates, rails, etc., are thus manufactured; and so on, until the furnace is emptied.

257. At the Lohe Works, in Germany, the charge is 164 kilog. Six heats are made in twelve hours. Every charge produces 7 to 8 blooms, weighing each 18 to 19 kilog. on an average.

The staff is made of fourteen persons:—

2 foremen.
4 puddlers.
2 hammermen.
1 man for the re-heating furnace.
5 assistants and helpers.
—
14

The waste of pig metal is 20 per cent.; 9 per cent. during puddling, 11 per cent. during re-heating.

1000 kilog. of steel require—

1.773 kilog. of mineral coal for puddling.
.342 " " " for re-heating.
———
2.115 kilogrammes.

IV.

Steel of Cementation.

258. The true term should be *iron of cementation*, because the product, obtained by the process we are about to describe, is far from being steel. It is nothing more than an iron which has been prepared to be transformed into steel, by introducing into its softened, but not fused molecules, atoms of carbon, which are suspended first, and afterwards dissolved in it.

259. Iron and carbon have a great tendency to unite, even when cold. Iron, left for some time in a mass of charcoal dust, will become hardened, and, by and by, may be transformed into steely iron, which is a kind of iron much sought for in certain works. Our iron-masters, who lose every year ten per cent. of their charcoal by turning to dust under their sheds, would do well to take notice of this phenomenon for producing a new kind of iron, which would find a ready market.

260. This has happened to two good men of Maine-et-Loire; one was an iron-master, the other an iron-dealer, both connected in business, with the same amount of intellect, and acting with that blind faith which characterizes the elect. The iron-merchant having bought from the iron-master a small quantity of bar iron, an idea came to his mind that

this iron would be improved if it was re-heated. The iron-master was induced to do the extra work, for which he was paid one franc the 100 kilog. The manufacturer, for the sake of economy, had the iron re-heated in charcoal dust of no value to him, the good man not suspecting that he was making a true cementation. As the expenses were covered by the extra price and the increase of weight, the operation was made thus: in a furnace out of use, a first layer of charcoal dust was made, and covered with a layer of iron, until the alternate layers of charcoal and iron had reached a certain height. Fire was then applied, and the whole allowed to burn entirely away. The good iron-master had no idea that he was making steel of cementation, and the worthy merchant was no more cognizant of the fact, although he did charge high prices to the blacksmiths, his clients, for this so-called refined iron. Very likely he would have remained in that ignorance, had not the author of this book revealed the mystery by breaking one of the bars before him. Yet, the revealer is not very certain that the candid merchant has been convinced.

261. The name of *cementation* is given to the operation by which two solid substances may be alloyed and dissolved into each other, without either of them being in a state of complete fusion.

In the cementation, iron and carbon are put in contact under a strong heat; but neither is melted.

We do not know how this penetration takes place; because, if there is solution, one at least of the two substances must be in the liquid state; and, in this case, it is difficult to believe that iron has acquired any fluidity.[1]

Fig. 4.

[1] Several explanations of a more or less speculative character have been given, in which the action of nitrogen plays a great part. Mr. Caron believes that nitrogen acts like a carrier, a kind of stevedore, in carrying molecules of carbon into the iron.

M. Frémy asserts, that without nitrogen, cementation could not take place.

There is always some air, and consequently nitrogen, in the charcoal used, and in the gases from the fireplace.—*Trans.*

262. The furnace employed for the manufacture of steel of cementation is a reverberatory furnace with a peculiar construction. Its length is about five metres; the fireplace is in the middle, and occupies the whole length, in order to render the heat uniform everywhere. Above the fireplace are built the flues, which sustain the cementing chests, and allow a free and equal passage for the flame. Upon these flues are two chests occupying the whole length of the furnace.

A free space is left for the passage of the flame around the two chests, which are built of large fire-bricks. Above, is an arch, which has the effect of concentrating the heat upon the chests. In the upper part of the arch, one opening is left for the exit of the smoke and the draft; another conical stack envelops the whole apparatus.

The furnace in use at Sheffield, and employed in some French steel works, differs slightly from the one we have just described; the opening left on the top of the arch for the exit of the smoke is replaced by two small chimneys built at each end of the furnace, one on the right side, the other on the left. The draft is more equal, and the heat is more concentrated under the arch; this disposition is therefore more advantageous.

263. The dimensions of the chests at Sheffield vary between eight and fifteen feet (2.44 to 4.57

metres) for the length, and between two and three feet (0.60 to 0.90 metre) for the width, in the clear.

The conical stack which surrounds them is 30 or 40 feet high (12 to 15 metres). The operation lasts six to eight days.

It is easy to understand that the dimensions of such a furnace are infinitely variable, according to the pieces which are to be cemented. Here are, however, the dimensions used in some works: the boxes are 4.25 metres in length, 0.60 to 0.90 metre in width and depth.

264. The iron which is to be converted into steel is generally a flat iron 6 millimetres thick.[1] The bars are cut in lengths of 3.90 to 4.00 metres, if they are to be placed in chests of 4.25 metres in length, in order to leave the room necessary for the expansion; otherwise, if the bars were to touch the sides of the chests, they would certainly shove out during the dilatation. To have everything ready in the chests, a layer of charcoal dust 0.07 metre thick is spread over the bottom; the bars of iron are laid upon this layer with a space between them; afterwards, another layer of charcoal dust 0.05 metre thick is added, another row of bars, and so on, until the chest is filled. Care has been taken that the space (0.07 metre) between the sides and the iron is

[1] Thicker iron is also used.—*Trans.*

filled with charcoal. The whole is covered with a layer of the same thickness as that of the bottom.

As it is important that the iron should be excluded as much as possible from contact with the oxygen of the air, the chests are covered with bricks hermetically luted. However, this method is unsatisfactory, because there are always some empty spaces in the chests, and the atmospheric air fills these spaces. In some works, the last layer of charcoal is covered with fine sand previously dried, and no other covering is employed. The sand follows the settling down, and steadily preserves the iron from contact with the air carried by the draft, and, at the same time, allows the vapors and carbonic gases to escape freely from the chest.

By the arrangement of iron and charcoal in a chest 0.60 metre high, we have:—

1 first layer of charcoal	0.07	metre.
8 intermediate layers of charcoal	0.40	"
10 layers of iron	0.06	"
1 upper layer of charcoal	0.07	"
Height of the chest	0.60	"

In the sides of the chests, small apertures are sometimes left, through which trial bars may be examined, in order to know if the operation is regular, and how it progresses. These trial bars are small bars or heavy iron wire; they are placed at different heights in the chests, and by them, it is possible to judge the degree of cementation of the remaining iron.

265. The apparatus being thus prepared, fire is begun in the furnace, first slowly, and afterwards increased gradually to its maximum, which is kept during the whole operation.

This operation lasts generally eight days, for furnaces in which fourteen tons of steel are treated at once.

When it is supposed that carbon has penetrated through the centre of the iron bars the most remote from the fire, the cementation is finished. After cooling, the charge is removed, and the chests are ready for another operation.

The iron which has undergone this treatment, has increased in weight from 0.005 to 0.0083. When taken out of the chests it is bloated in many places, and bursted in some others. These bloats, or rather these blisters, are the cause of its name *blistered steel.*

This steel requires to be reheated and hammered or laminated before it is sold; it is then called *tilted steel* or *drawn steel.* The preference is given, with reason, to the hammered or tilted kind, although it does not present so good an appearance as the other.

266. Any kind of furnace is suitable for making steel of cementation; the ordinary reverberatory furnace will answer perfectly well for this manufacture, provided that the chests are properly built, and that the flame plays all around them.

At present, large furnaces are constructed for cementing pieces of forged iron of large size, such as rails, wheel tires, anvils, etc. In this fabrication, the object sought for is only the cementation of a small portion of these pieces; therefore, the cementation is ended as soon as it is thought that the carburization has penetrated deeply enough. The degree of cementation is also proportioned to the various uses of these pieces. Two advantages are thus obtained: one is a partial *acieration* (cementation) of the piece; the other is an intimate union of metals having a different density.[1]

V.

Cast Steel.

267. Natural steel, raw steel, puddled steel, cemented steel, and all those produced by the Bessemer, Uchatius, Taylor, and other processes, have little homogeneousness, and are the result of a solution of a greater or less quantity of carbon in the definite alloy.

In order to reduce these mixtures to a definite alloy or steel, the carbon must undergo an operation by which its quantity is regulated. This will be effected by allowing the metal to rest some time in a perfectly fluid state, at the bottom of the crucible,

[1] It has been said that iron bars holding sulphur, may get rid of this impurity by the cementation process. In this case, a bisulphide of carbon would be formed, and expelled on account of its great volatility. We do not know if experiments have been made in confirmation. *If true*, a good use would be found for iron bars whose main impurity is sulphur.—*Trans.*

at a high temperature, and out of contact with the air.

268. Therefore, the crucibles or "pots" which receive the imperfect steel to be smelted, must be able to resist a very high temperature, and that for a certain length of time. They are the most important tools of the steel manufacturer, whose profession does not require a great scientific knowledge.

However, the construction and the moulding of the crucibles or pots has been, until now, left to the rule of thumb and to the convenience of workmen; the proprietors of furnaces, out of good-will, have been dependent on ignorant moulders for what forms over half the expense of their industry. In this respect, no metallurgic district is so far behind, as St. Etienne and its neighborhood, where, however, is to be found more scientific knowledge than in any other part of the world.

Lamenting this state of things, I thought it would be useful to describe minutely the manufacture of crucibles or "pots;" the more so, as such description is not to be found in works of this kind, and has been neglected, as if it were of no moment.

269. The crucibles must be made of materials as refractory as possible, according to the place of manufacture, and the price of clay. The cost of fabrication will be the guide on this subject.

270. The refractory clays vary greatly, and are distributed everywhere; there are few countries where they cannot be found. The geological nature of the ground will often point out the quality of the clay.

All silicates with a basis of alumina are adapted to the manufacture of refractory materials. But it is very important that they should not contain potassa, soda, lime, alkalies, or metallic oxides, at least in notable quantity.

In the primeval rocks, amid the debris of gneiss, deposits of kaolin occur which is highly refractory.

In the rocks of transition, many kinds of suitable silicate of alumina are found. Often, these rocks contain deposits of graphite which, when abundant, is the best refractory material in use. Such silicates occur also in the secondary rocks; but it would be useless to search for them in the upper formations. Indeed the clay of tertiary rocks holding lime is useless, on this account, for our wants.

The refractory clays generally contain 50 per cent. of silica, 33 of alumina, and the balance water. The presence of alkalies or metallic oxides would injure their quality.

271. Many earths may be variously mixed, thus acquiring refractory qualities. One might unite a very white river sand with a sufficient proportion of alumina,[1] to give it some cohesion and resistance.

[1] By alumina the author certainly means "clay" or silicate of alumina.—*Translator.*

One-fourth of white or slightly blue alumina and three-fourths of ground white sand give an excellent paste, which well resists the fire.

The English fire-bricks are made the same, only the sand of the mixture is fine and has not been ground. Coming out of the kilns, they have a slight pink color, and their cohesion is so feeble, that the edges will easily crumble under the fingers. However, they are justly celebrated for the building of furnaces.

Nevertheless, it rarely happens that the pots for smelting steel will resist during a great number of operations: being put in close contact with the fuel, the ashes, carried away by the draft, will stick to their outer surface, and will cause a rapid vitrification.

The clays which have been chosen according to the above indications, must undergo several operations before moulding.

It will be good to leave them exposed to the air and rain for a certain length of time; water will wash out a large portion of the iron oxides, which they always contain in a greater or less quantity.

When the clays are to be used, they must be ground, sifted as fine as possible, and washed with great care. Afterwards, they are mixed, pressed, drawn, and corrugated until they have acquired the proper plasticity. When they are firm enough to be moulded, the workman kneads them again with a pestle or a mallet, and forms lumps large enough for the size of crucibles. Care must be taken that these lumps should be a little more than what is necessary, but never less; because an addition of clay, when making the pot, would destroy its solidity.

272. The moulds used for makingthe smelting pots for steel, are of brass; cast iron would be too heavy, and iron too expensive. The annexed figure shows their form. Their size varies with the quantity of steel to smelt. Generally the depth is 0.90 metre, the diameter in the clear 0.16 metre, and the thickness 0.02 metre for the sides, and 0.03 metre for the bottom; they have two projecting handles for lifting. The bottom, which is concave, is perforated with a hole of about 0.05 metre diameter, and receives a movable bottom perforated with a hole 0.02 to 0.025 metre diameter. This movable bottom has a convex surface, which corresponds as nearly as possible with the bottom of the mould, and a flat surface upon which the pot rests.

Fig. 5.

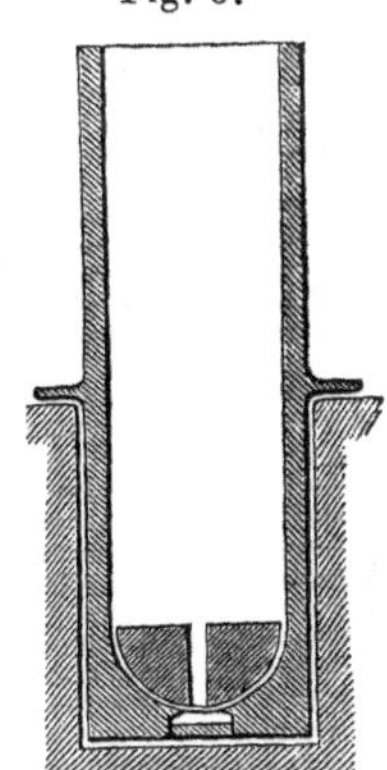

The mould stands firmly in a hole made in the floor, and room is left to allow the workman to reach the handles. In order to give more solidity and correctness, the bottom and sides of the hole are lined with strong pieces of wood. At the bottom is also a perforation of 0.02 to 0.025 metre bore, which corresponds exactly to that of the movable bottom. It is used for passing the centre spindle of the plugs.

The mould thus prepared is smeared with oil, and the workman throws into it a cylindrical lump of clay, which he presses firmly with a kind of pestle (see figure 6). Afterwards, a centring board made of wood, lined with iron, and with a hole equal to that of the movable bottom, is inserted into the mould. Through the hole of this centring board a

Fig. 6.

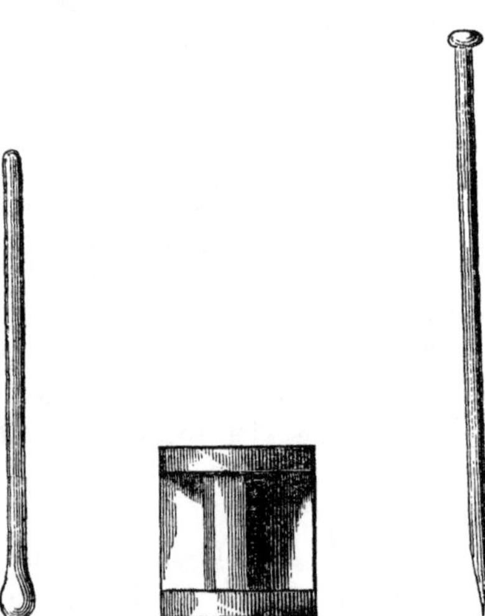

spindle of iron is passed and pressed downwards, until it goes, through the clay and the movable bottom, as far as the hole bored in the basis which supports the moulds.

When the spindle and the centring board are withdrawn, the crucible is centred. The workman then takes a first plug, well oiled, and adjusts it to the spindle which rests in the hole previously made. With a mallet, beginning with gentle blows first, and heavy ones afterwards, he drives the entire plug down into the mould.

The plugs are made of wood with a mounting of iron at the top. The iron spindle is round and pointed at the lower part, round and with a head at the upper one, but square where it passes through

Fig. 7.

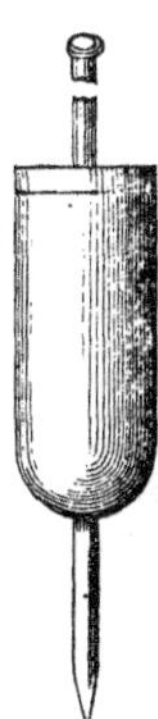

the wood. Near the head, a hole is bored, through which a pin is inserted, which allows the workman to give a screwing motion to the plug. The clay becomes loosened, first from the mould, and soon after from the plug; the workman then takes another oiled plug, which differs from the former only by being longer (0.70 metre.)

This new plug is driven in until its upper part is level with the top of the mould, and is withdrawn. Part of the clay having risen 0.10 metre above the mould, the workman gives it a truncated conical form, by means of two tin-plate moulds. But, previous to this, the hole left at the bottom of the crucible is scraped with a tool here represented, in

Fig. 8.

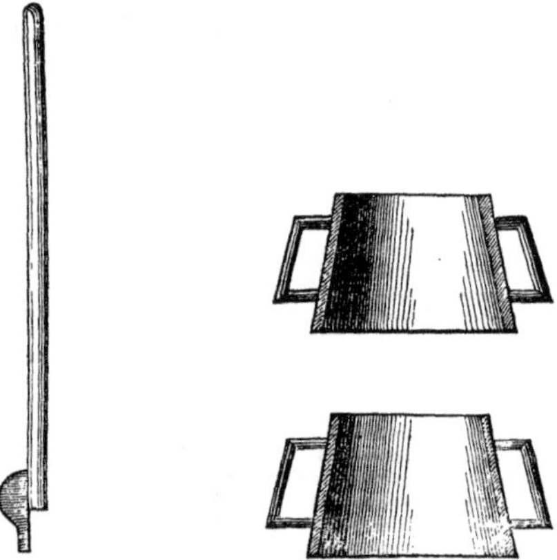

order to extract all the clay which has been impregnated with the oil used for smearing the spindle and the plug. Next, a small lump of clay, placed at the end of a stick, is pressed into the opening, and the whole firmly united with another piece of wood, similar in shape to the bottom of the crucible.

The pot, with the bottom hole closed, and the top bent as we have said, has the appearance shown in the next figure. The next operation is to extract it from the mould. To do this, the workman takes the mould and carries it to another hole dug in the floor, in the centre of which is a vertical rod of iron, terminated by a flat head of 0.05 metre diameter. This

Fig. 9.

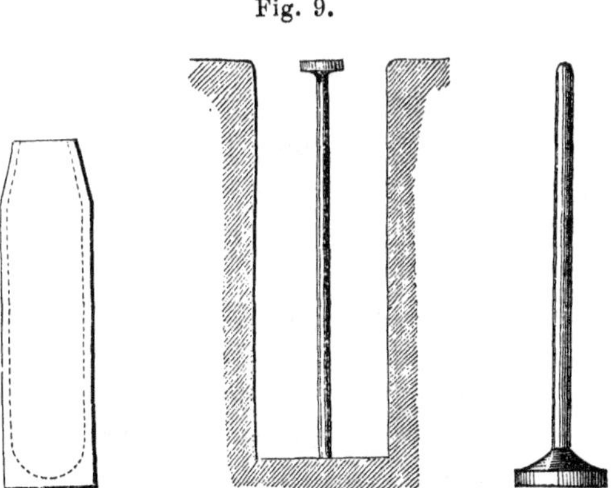

head passes through an aperture left at the bottom of the mould, and raises the movable bottom with the pot on top, while the mould is lowered down the floor hole. The workman then takes the pot, pulling gently to free it from any of the clay which has got into the centre hole, and carries it to the board, where it is left to dry awhile in the air.

273. The drying is finished in a drying-room, where the temperature is kept at about 30° C. (86° F.), and lasts several days, until all moisture is expelled. When the pots are to be used, they are put into the smelting furnace, where the temperature is increased gradually up to a cherry-red heat; at this point, they are ready to receive a charge.

274. The annealing of the pots presents great difficulties, and requires a great deal of care: the fire must be regular; a sudden increase will make them crack; once properly heated, they will easily bear the changes of heat and cold to which they are subjected. Plumbago crucibles, particularly, after a powerful heating, may be dipped into water without being injured.

275. In many works, the upper part of the crucibles is straight, without the conical form adopted by others. Their manufacture is rather simplified; one plug will be sufficient for the moulding; thus, three tools and three manipulations are dispensed with. However, the conical form has some advantages, as a greater facility in covering and uncovering during the fusion, mostly when several pots are put in one furnace.

276. In the place where the pots are moulded, and with the same clay, small flat disks are made, to be used as stands and lids for the pots, with which they

are dried and burnt. These stands or supports for the crucible protect its bottom, and, at the same time, raise it to where the fire is the most powerful.

277. In a day of ten hours' work, a workman can make thirty pots, preparing the clay besides. In some steel works, where improved methods are in use, as many as six successive fusions are performed in the same pot.

Many more might be effected in plumbago crucibles; but their cost is greater and the materials are difficult to obtain.

278. Graphite or plumbago is a carbide of iron very rich in carbon; its two principal characters are to leave a black streak on paper, and to be infusible.

Graphite occurs in the crystalline rocks, where it forms veins, or irregular pockets, like chaplets. At Borrowdale the pockets are somewhat considerable; at Cabo-de-Penas, in Asturia, the veins are of a small depth. It occurs also at Pontivy, in Brittany; at Marbella, in Andalusia; in Bavaria, Siberia, America, at the Cape of Good Hope, &c.

279. The graphite employed in the manufacture of pencils contains sometimes as much as 96 per cent. of carbon; as, for instance, the Borrowdale graphite. At Sheffield, the graphite in use is properly a refractory clay with a small percentage of plumbago, scarcely 9 per cent. However, some of

these graphites may be mixed with over one-half of silicate of alumina (clay), and resist well the most ardent fire, during fifteen successive charges of enamels with a basis of lead.

280. The furnaces employed for the fusion of steel are similar to those used in brass foundries, or in laboratories; they are air furnaces, made with a rectangular hearth having at the bottom a grate, upon which rest the crucible and the fuel.

281. These small furnaces are 0.50 metre long, 0.35 to 0.40 metre wide, and 1 metre deep above

Fig. 10.

the grate.[1] On top is a movable cover, which allows the putting in of the crucibles, the fuel, the charge, and, at the same time, enables the workman to see how the operation is progressing. At the upper part of the wall side is a flue communicating with the chimney or stack, which produces the draft. Underneath the grate, and in front, is an opening for the access of air and the reception of ashes.

Such is the disposition of casting furnaces in many steel works. In England, their construction has been slightly modified, in order to secure more facility in the working.

282. The Sheffield furnaces are made on the same principle; but the floor of the workshop is level with the top of the furnace, and the covers are fixed by hinges to the wall at the back. A chain, with a counterweight, passes into a pulley-block fixed in the ceiling, and the other end is attached to the cover, which may be thus kept open, when wanted.

Access to the ash-pits is by the under floor or cellar. These small furnaces are built contiguously to each other. A horizontal flue is used for the draft, and is connected with a central stack which does not require to be higher than 6 or 7 metres.

Each furnace receives two crucibles which are put upon the stands already mentioned. Burning coals are added, and the temperature is gradually increased

[1] These dimensions are very variable.—*Trans.*

by small additions of fuel, until the pots are red. The charge of steel of cementation is then put in.

283. Generally, in France, the pieces of natural steel or of cemented steel are thrown into the pots; this might take corners off, or even break the bottom. At Sheffield, a long sheet iron funnel is introduced into the pots, and the pieces of steel to be melted are thrown into it. The use of this *charger* is a good idea, which we recommend to our countrymen.

284. The fuel used at the beginning is pit coal; coke is afterwards employed in pieces as large as the first. The coke must be very dense and compact; that made in furnaces is better than that made in the open air. A very dense and compact coke will last longer, and will not necessitate the addition of a fresh quantity during the operation. Indeed, good fuel and in sufficient quantity, will last long enough for a complete fusion. The addition of new fuel might be injurious to the pots, and it would be difficult to spread it equally; the fusion might thus be irregular.

Some manufacturers, instead of coke, use pit coal, the flame of which surrounds the casting pots. The furnace is then different, and has the appearance of a reverberatory furnace with a fireplace for the fuel, which is no longer in contact with the pots. It would be really advantageous to employ pit coal, which will produce 5000 units of heat instead of

4900 for coke, as we have shown (49); but the difficulty, until now, seems to have been in combining a good draft with a complete surrounding of the crucibles by a powerful and constant flame. Experiments made on a large scale at St. Etienne and at Chambon, had to be abandoned.[1]

285. The steel of cementation, broken into small pieces, is introduced into the pots through the funnel or "charger" here represented. This charger is so constructed that it will go into the crucible, which will be thus protected against breaking by the pieces of steel thrown into it. When the charge is made, the pot is covered with a lid, the necessary quantity of fresh fuel is added, and the process is allowed to go on until the steel is completely fused, which takes place after a length of time learned by practice.

Fig. 11.

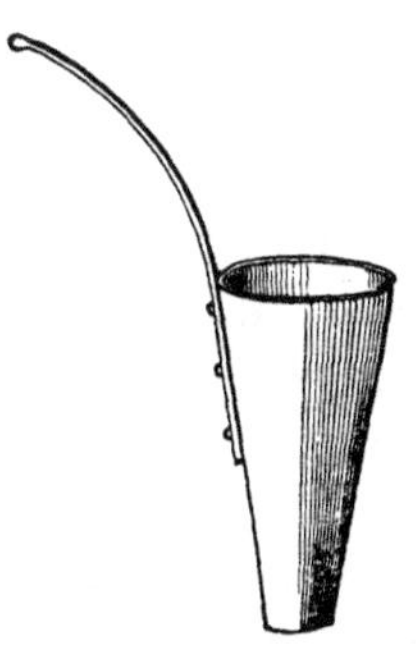

However, the length of the operation varies with the size of the crucibles; in some English works it is done in two hours; in some, six hours are required; in others, three hours only are necessary.

It being important that atmospheric air should not penetrate into the pots, they are properly

[1] Siemen's gas furnaces are well spoken of.—*Trans.*

covered. For this purpose, the top of the pot has been ground over a flat stone, so that the lid or cover fits it exactly. In many steel works, the clay for the lids is not very refractory, in order that, under the intense heat, it may become somewhat vitrified and adhere to the top of the pots, thus completely preventing the entrance of air.

During the operation, the workman watches the fire continually, so as to add fuel if that in the furnace is sinking; but this rarely occurs, if the fuel has been broken into pieces about the size of an egg. Sinking generally takes place when the pieces are of unequal size. When it is thought the operation is complete, the furnace is uncovered and left open for a few minutes, till the lids of the pots have cooled somewhat, that they may be easily removed. By waiting too long, they would harden so as to make it difficult to separate them; it is sufficient if they have become hard enough to resist the tool. The workman then takes hold of the uncovered pot with the lifting tongs, and removes it from the furnace.

A small quantity of slag, which swims on top of the molten steel, is withdrawn with an iron bar or *flux stick;* afterwards, the contents are poured into an iron ingot mould having within an octagonal form, and which has been previously smeared with some clay or plumbago, in order to facilitate the separation of the steel ingot after cooling.

286. When the steel is poured into the mould, it

often happens that it ascends and runs over the top. To avoid this, the mould is immediately covered. Great dexterity is necessary in pouring. In some works, immediately after pouring, the aperture of the mould is closed with an iron stopper, upon which the workman strikes, gently at first, more heavily afterwards, according as the cooling progresses.

By cooling, steel contracts, and a pipe hole may be produced; this must be prevented as much as possible. The contraction, taking place from the periphery to the centre, principally when the pouring in has been too rapid, and the air had not the time to escape, will leave in the centre an opening of the size of a small finger in the whole length of the ingot.

To obviate this difficulty, the molten steel should be directed towards the middle of the mould, with out touching the sides, and the pouring should be done slowly.

When the steel is cold, the mould is opened and the ingot taken out; but it must undergo a new operation, in order to acquire the fine and close grain of the commercial cast steel. Directly from the mould, the grain of the steel is coarsely crystallized and somewhat similar to that of fine metal; it must be refined, *i. e.*, reheated and drawn under the hammer to a bar.

The next figure is a vertical section of an ingot mould; this is made of two parts, which join her-

Fig. 12.

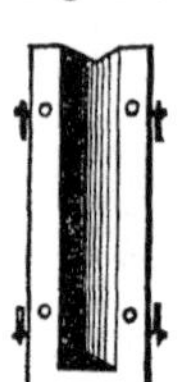

metically and are kept together by means of strong movable clamps and iron rings. The mould must be larger than what is required for the quantity of steel to be poured in. This quantity is generally 15 to 20 kilogrammes. However, at present, larger ingots are manufactured by pouring into the same mould the product of several crucibles. Pieces of steel of over 100 kilogrammes are often cast in this way.

287. There was exhibited at the exhibition of industry a block of steel of 6 tons; such a weight indicates some other system than fusion in crucibles not holding more than 20 kilogrammes. These fine samples, perfect in quality, were from Prussian steel works;[1] they must have required special casting apparatus, but on known principles. Intended for exhibition, such masterpieces of manufacture are too costly to be applied to the arts, except in certain particular cases. Therefore, being without interest as regards industrial uses, we shall not dwell any longer on this point, which is beyond the limits of this work.

288. Manganese is a powerful auxiliary in the manufacture of cast steel, and its use is becoming more and more general. We have explained the

[1] At Krupp's works, ingots of 40 tons have lately been cast.—*Trans.*

effect of such an addition, which regulates the quantity of carbon in cast steel.

289. Indeed, steels of cementation will not present, even were they taken from the same cementing chest, a complete uniformity in their composition; the carbon penetrates the bars only partially, and the pieces of steel put into the crucible are heterogeneous, and with an irregular percentage of carbon. When it is ascertained that the steel of cementation is wanting in carbon, it is advantageous to throw some ground charcoal over the charge; but then, it is necessary to have a guide, a substance regulating the proportions. Manganese possesses this property. In the bottom of the crucible certain reactions take place, which are a complete mystery to the manufacturer, and have a great analogy with those of the production of natural steel. Cast steel, in order to become a chemical alloy (167), must remain some time in the crucible, in a molten state. If fusion alone were needed, it could be effected in one hour; but in this case, it seems that steel requires time to perfect itself. It is certain that changes occur in the molten mass, standing perfectly still, without any mechanical stirring, because the coarse crystals of steel of cementation are transformed into the close, fine, and compact crystals of cast steel.

It might not be out of place to speak, here, of an invention by a manufacturer, Mr. Ballefin, for heating the crucibles in a furnace where the air is forced through, and where a remarkable economy of fuel is effected. But this process having been monopolized by a company, and not bearing directly on the manufacture of steel, we may omit it, regretting, however, that it is not within the reach of all cast steel manufacturers: then our industry would have been able to compete advantageously with the Sheffield producers.

VI.

Wootz.

290. *Wootz* is the name of a certain kind of steel manufactured in India, and which appears to have been known from time immemorial, it being a historic fact that Porus gave 30 pounds of it to Alexander.

291. This steel is made by the natives from a magnetic ore, very rich, and in which silica or quartz is the only impurity. Its composition is—

Iron	37.67
Oxygen	14.33
Quartz or silica . . .	48.00
	100.00

At a glance, it is evident that the steel from such ore will contain a certain quantity of silicon, but no aluminium, as is asserted by some metallurgists.

This mineral occurs in great abundance in the district of Salem, where a great deal of steel is manufactured. It forms large hills, and is extracted from the surface. It undergoes a preliminary preparation which consists of a stamping, after which the foreign matters are mechanically removed.

292. The furnaces used for the manufacture of Indian steel vary much according to their locality. At Salem, their form is conical, and the height is not over 3 to 4 feet. The bellows are made of two dog-skins fitted to a bamboo tube, itself tipped with a clay tuyere. The ore is put upon a thick layer of charcoal, not otherwise prepared, in the fashion of the Catalan forge. After four hours of blast, the reduction into steel is effected, and the liquid metal obtained is allowed to cool. However, before the cooling is complete, and while the steel remains red, it is cut into pieces with a hatchet, and delivered to the blacksmith.

293. This steel is far from being homogeneous, and is very much like our natural steel. If it were to remain so, it would not possess the just celebrity acquired by the Wootz steel.

294. The pieces of the crude steel are drawn into bars with the hammer, at a high temperature; then these bars are cut into small pieces, which are put into a crucible with some dry wood of *Cassia auriculata,* and some green leaves of *Asclepias gigantea*. The charge is about one pound of steel. In order to

prevent any access of air, a lid is forced into the crucible, all the joints are perfectly luted with clay and allowed to dry. Twenty crucibles thus prepared are piled up in the same furnace; the whole is covered with charcoal, and fire is immediately applied. The fusion lasts two hours, and an excellent steel is produced.

295. The mode of working by the natives is so imperfect, that, out of 64 per cent. of iron in the ore assorted and calcined, only 15 per cent. is extracted.

296. This method, aside from its imperfections, has a great analogy with the processes followed in Europe. After all it is cast steel, from natural steel, perfectly refined.

297. The process presents differences according to locality. Sometimes the furnace where the ore is reduced is 4 to 5 feet high, and so conical that the upper part is one foot diameter, while the lower part is five feet. This shape is certainly for a better concentration of the heat. In some districts, the furnace is entirely made of fire clay, and is built in a few hours. The front part has an opening about one foot high, shut up with clay, and destroyed at every operation. The bellows are made of the skin of a goat, taken from the animal, without any longitudinal incision. The legs are sewn up, and the neck is tied around a bamboo tube. The incision

at the tail end has its edges straightened by pieces of bamboo, thus making a valve which can be opened and shut. Handles of wood or leather allow the man to work them up and down. Two such bellows are required, and by alternately pressing them, a steady blast is kept up. The bamboo tubes are inserted into other clay tubes, which are the real tuyeres of the furnace.

This is filled with charcoal, and some light burning material being put in front of the tuyere, the combustion soon becomes general. At this moment the ore, previously moistened to prevent its falling through the fuel, is put on top of the charcoal, and the whole is covered with enough fuel to last three or four hours under the action of the blast. Immediately after the operation is completed, the bellows are stopped, the front of the furnace is broken, and with a pair of tongs the metallic lump is extracted from the hearth.

In other places, the furnaces are also four to five feet high, but they are more narrow. Some have the shape of a truncated cone, whose base is two feet in diameter, while the top is one foot.

298. The Wootz steel is sold in India in the shape of round disks, whose diameter is about 0.126 to 0.127 metre, and thickness 0.025 to 0.026 metre. Each weighs about two pounds; the color of the outside is black, and the surface is smooth. The texture is regular, the hardness extreme, and the

heaviest hand hammer will leave no impression on it.

VII.

New Processes.

299. All the methods for manufacturing steel in an industrial way may be reduced to five:—

1. Reduction of the ore and carburization of the iron, or the direct process;

2. Partial decarburization of pig iron in a finery fire;

3. Partial decarburization of pig iron in a puddling furnace;

4. Cementation of iron;

5. Fusion, by which homogeneousness is given to any of these steels.

Therefore, cast steel, or perfect steel (the true definite alloy), requires two distinct operations:—

The carburization or cementation of iron, and the partial decarburization of pig iron.

The fusion or the transformation of the carbide into an alloy with definite proportions.

300. Metallurgy has yet, as regards the chemical part of the production of the metal, an important improvement to make; it is to reduce to one operation, the two at present required. Otherwise, it is to produce the definite alloy, cast steel, by the direct union of iron and carbon.

301. If we have well understood the various processes known and in use, of which we have spoken, the chemical or definite steel, which is the true steel, is obtained in the following way: When refining pig iron in a low furnace, or in a puddling furnace, the metal is smelted rapidly, and allowed to stand a certain length of time under a layer of fluid slags, shut up from the air. There it is converted into steel, either by losing part of its carbon, or by a change in the union of carbon with iron. The presence of oxide of manganese will help this reaction.

In the fusion of steel the same phenomenon occurs; perfect quietness, without contact with the air, more intimate combination, and the regulating action of manganese.

We do not speak now of the two other modes of making steel, *i. e.*, the cementation of iron, or that of the ore (direct process), as neither affords regularity or certainty. The former succeeds, only because the cementation absorbs an excess of carbon; the latter has so much uncertainty in it that the workman is never sure to produce steel, and often he will make ductile iron instead.

Our deductions from the practical modes of manufacturing, corroborate the inferences we have drawn from theory at the beginning, *i. e.*, that iron and carbon alone are sufficient to make steel, and that manganese is useful only to regulate the proportion of carbon. Silicon, aluminium, and other earthy metals, may be dispensed with, and are only accidentally to be found in the carburized metal.

It will be noticed, that in all these processes, there is no exact, mathematical proportion, and that the incertitude and the approximation which predominate in them leave too much to chance.

302. In one of these processes, Mr. W. E. Newton[1] employs iron ore itself, instead of iron. In a cementing furnace, alternating layers of ore and charcoal are piled up with the flux the ore may require. The furnace is kept during forty-eight hours at a white heat. According to the inventor, the iron agglutinates into irregular sheets, and the slags are mechanically removed. Afterwards, the metal is cast, drawn, and worked for springs. We confess that this result seems to us very uncertain; it is contrary to the theory of the reduction of iron, and we are afraid the author is mistaken. However, Mr. Newton is the only inventor who, among the so-called discoverers springing up everywhere, has had the daring to take in hand the direct use of the ore.

303. Several metallurgists, among them Mr. Crace Calvert,[2] M. Fontaine, of Paris, Mr. Martien, of New Jersey,[3] Mr. Tilghman, of Philadelphia,[4] have obtained patents for the employment of chloride of sodium, and even chlorine, in the manufacture of steel.

[1] Patent of 1855.
[2] Patent of 1851.
[3] Patent of 1856.
[4] Patent of 1856.

304. At the close of the last century, David Mushet[1] had used common salt in the metallurgy of iron; Samuel Rodgers[2] had recommended its employment in 1819 by the iron works of Glamorganshire.

305. Chloride of sodium acts only by its alkali, which is an excellent flux for separating silica from iron. Therefore, in certain cases, it may be useful in the metallurgy of iron; but in the manufacture of steel it does not act as chloride, and even less by its chlorine which has been proposed by Mr. Martien. When Mr. Brooman obtained a patent (1854) for the use of manganese and chloride of sodium, he must have relied upon the first substance for the manufacture of steel. The other substances which, in 1856, he proposed to add to the two former seem to have been picked up without discrimination, as they have little, if any, action upon carburized iron.

306. Recently (1856), Mr. Robert Mushet, son of the celebrated metallurgist of Clyde iron works, obtained a patent for the addition of manganese which he pulverizes and throws upon the molten metal. In 1856 this distinguished manufacturer had taken not less than eight patents, all relating more or less to the use of manganese.

Some manufacturers have tried to unite pig iron with ductile iron, in order to unite with the latter the excess of carbon of the former, and thus to make a

[1] Papers on Iron and Steel, p. 133, and following.

[2] S. Rodgers' Letters, unpublished.

kind of commercial steel which is not the chemical alloy.

307. Messrs. Price and Nicholson have proposed to cast together fine metal and wrought iron,[1] and Mr. G. Brown[2] has alloyed charcoal pig metal with iron made from the same metal.

308. Mr. Manory goes further;[3] he not only alloys white metal with iron broken into small pieces, but he adds to the molten mass oxides of iron, calcium, sodium, and potassium.

309. In 1854 Mr. Sterling had an idea more simple and certain: it was to add to the raw metal, smelted in a crucible or in a reverberatory furnace, progressive quantities of oxide of iron; as long as it was necessary to improve the steel. The oxide used was as much as possible a magnetic ore. Besides, and in order to give body and hardness to the carburized iron, some oxides of tin or zinc were added.

Mr. Uchatius, whose process has attracted a great deal of attention in England, and to which on that account we will give a special chapter, employs also the oxide of iron which reacts upon the pig iron which is very finely granulated. The patent of Mr. Uchatius was taken out in 1855, and that of Mr. Sterling in 1854.

[1] Patent of 1855. [2] Patent of 1856. [3] Patent of 1856.

The oxide of iron used for decarburizing pig iron was naturally to lead to the idea of employing the oxygen of air or of steam to produce similar results. Mr. Martien was the first to think of forcing a blast through cast iron in a perfect state of fluidity.

310. It is certainly to the idea of Mr. Martien that the origin of the invention of Mr. Bessemer is due. This manufacturer, after having tried for a length of time the alloying of pig metal and iron according to the process of Messrs. Price and Nicholson, and after having originated two kinds of cupolas where the smelted pig iron passed alternately from one to the other, at last succeeded in decarburizing pig metal by a violent blast forced through its particles in the molten state. As an entire chapter is devoted to the Bessemer process, we shall not here explain it.

The various processes we have just described do not require more explanation. They have not all been sanctioned by experience, and most of them have remained in the state of theory. We have ignored many of them, as being without any industrial value. The four following processes, although yet remaining in the state of experiment, seem to us each worthy of a special chapter.

Chenot Process.

311. In the ordinary blast furnace where pig iron is produced, there are two distinct operations, at two different heights of the stack. The iron ore,

which is a compound of iron, oxygen, and earthy matters, is reduced in the upper parts of the furnace, after somewhat complex reactions.

1. At a certain height, the ore comes in contact with carbonic oxide (oxide of carbon), which unites with its oxygen, and escapes at the mouth in the state of carbonic acid.

Fig. 13.

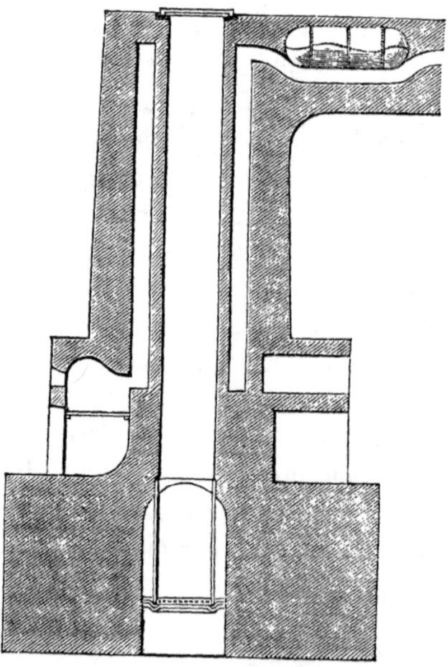

2. The earths which accompany it become separated and fall to the lower part of the furnace, where they are transformed into slags or cinders.

3. The iron is reduced and remains pure. The height, where these reactions occur, is termed the zone of reduction. The heat is intense. At this point let us see what happens.

The carbonic acid escaping at the top of the furnace, the iron and the slag remain, which, on account of their specific gravity, fall to the lower part of the furnace, where the temperature is greater, and where they undergo new reactions.

312. Every one, at this stage of the operation, will think it would be more advantageous to extract the perfectly pure iron, instead of allowing it to fall among the incandescent coals, where it is transformed into a carbide or pig iron.

313. Such was the idea of Mr. Adrien Chenot, who thought of stopping the operation just when the ore had been converted into pure iron. An intense heat reigns in the greater part of the furnace, five metres above the zone of reduction, and ten metres underneath. Why should such a heat be kept ten metres under the point where the pure iron is obtained? We readily understand that when pig iron is wanted, such a heat is maintained underneath in order that the iron shall be converted into pig metal in a carburizing atmosphere. But, when ductile iron is wanted, we think it is useless and costly to keep up the combustion of the fuel, when the product sought for is already obtained.

314. Following these principles, Mr. Chenot, instead of heating the furnace underneath the boshes, *i. e.*, the lower part, has directed the highest temperature to be applied at the zone of reduction. The combustion ends there. The pure iron thus produced descends gradually into cold boshes, where it cannot undergo any new reaction, and where it is found in spongy masses mixed with the earths.

The only operation which remains is a mechanical separation of the iron from the earths. This is effected by powerful magnets, which, being presented to the cold and pulverized residua, separate the iron in a state of perfect purity.

315. This powder of pure iron being submitted afterwards to an enormous pressure, which reaches above 700 atmospheres, has its atoms so strongly united that it acquires the density of iron itself, and may be drawn into bars, and undergo all the operations of a forge.

316. This compressed sponge is used by Mr. Chenot for cementing the iron, and converting it into steel. It has been found, by experiment, that it will absorb its own volume of liquid; therefore, it is sufficient to dip it into an oleaginous liquid, such as coal tar,[1] in order to produce a true carbide—not

[1] Wood tar is considered preferable, as it is more free from sulphur than coal tar.—*Trans.*

by combination, but by mixture. The metallic mass thus impregnated is put into pots, and cast in the usual way.

317. This new metallurgy of steel, where all the operations can be made without heat, from the reduction of the iron to the fusion of steel, will certainly revolutionize the metallurgy of iron. It is the greatest thought that we have had for a long time in applied science.[1]

Bessemer Process.

318. The process of Mr. Henry Bessemer is a decarburization of pig iron by a powerful blast of air, whose divided molecules pass through the carburized iron in a liquid state. Therefore, the oxygen of the air being in intimate contact with the carbon of pig iron, takes of it just what is necessary to leave the definite alloy.

319. Mr. J. G. Martien (310) had discovered that

[1] We do not desire to disparage the efforts which may be made in that direction, but some facts in working the Chenot process will show the great difficulties to overcome before all sanguine expectations can be realized. The ore must be perfectly pure, free from earths, and the reduction complete. If all these conditions are not fulfilled, part of the earths and of the unreduced ore will remain in the sponge of iron, and will not be completely separated by the magnet. Besides, all the necessary extra precautions are costly. The Chenot process, truly remarkable by the simplicity of its principles, has not yet found its way clearly into practice.—*Trans.*

by passing a blast of air through molten pig metal, not only the carbon of the raw metal was destroyed, but also that the temperature remained high enough for keeping the iron in the liquid state and for casting it. This discovery of enormous importance had however remained in the state of theory, and would have made but a feeble sensation, had not Mr. Bessemer undertaken to apply it.

320. In his former experiments, Mr. Bessemer was running molten pig iron from a blast furnace or from a cupola into another cupola, at the bottom of which several tuyeres were giving the necessary blast. If the blast is shut off just at the proper time, steel is produced; if the blast is allowed to act long enough for completely decarburizing the metal, ductile iron is made. The great difficulty in practice is to ascertain the proper time for stopping the blast. Generally, the pig metal is over decarburized; and to make up for the deficiency of carbon in the steel, a small quantity of molten *spiegeleisen*[1] is added to the metal in the converter. A few minutes more of blast will thoroughly mix the whole mass, which is then ready to be cast into ingots.

The first idea of using spiegeleisen is due to Mr. Robert Mushet. This metal acts by its carbon and

[1] Spiegeleisen (*mirror iron*) is the German name of a kind of pig iron very rich in carbon (5 per cent.) and manganese (4 per cent.).—*Trans.*

its manganese at the same time. From its employment dates the practical turn of the Bessemer process.

In this process, and in all others where the oxygen of the air acts alone, very pure pig metal is needed for the manufacture of pure steel or iron; silicon and carbon will be removed, but sulphur and phosphorus will remain in the manufactured product if they were already in the raw iron. For removing these latter impurities several substances have been proposed and experimented upon; time and practice will determine their value.

Fig. 14.

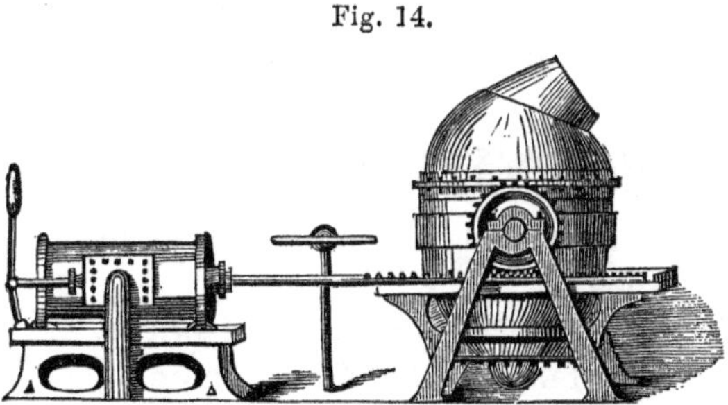

Converter with the engine.

321. The cupola of the former experiments has given place to the above apparatus. The converting vessel or *converter*, which we will describe more completely hereafter, revolves on two trunnions. One of them is hollow and is connected by a coupling box with the blowing machine, the blast passing through a curved pipe along the lower part of the converter

and terminating in a metallic box beneath the apparatus. The other bears a strong pinion, to which a revolving motion is given by a rack at the end of the piston-rod of a double-acting, water-pressure engine.

322. The converter itself is an ellipsoidal vessel, made of strong wrought-iron plate. The lower and upper parts are bolted together. On the top is an oblique mouth for receiving the charge of metal, and for the escape of gases, &c. At the bottom, a metallic box receives the blast and divides it through the tuyeres, five, six, or seven in number. The trunnions are fixed upon a large wrought-iron belt, about midway of the apparatus. The inside lining must be made very carefully; the refractory clay, strongly beaten into it, is mixed with a certain quantity of quartz (*ganister*), or ground fire-brick free from scoriæ (*chamotte* or *cement*).

The hole of the tuyere is also made of fire-bricks, with all the joints carefully luted. When the lining is dry, a charcoal fire is built in it, and all cracks closed. Afterwards, a stronger fire is built, some blast is given, and the interior receives a glazing of common salt.

The ashes having being removed, the apparatus is ready for working. The converters are made to receive from three to five tons of molten pig iron, which should, however, occupy only a small place in it; the reaction and the boiling are so violent,

Fig. 15.

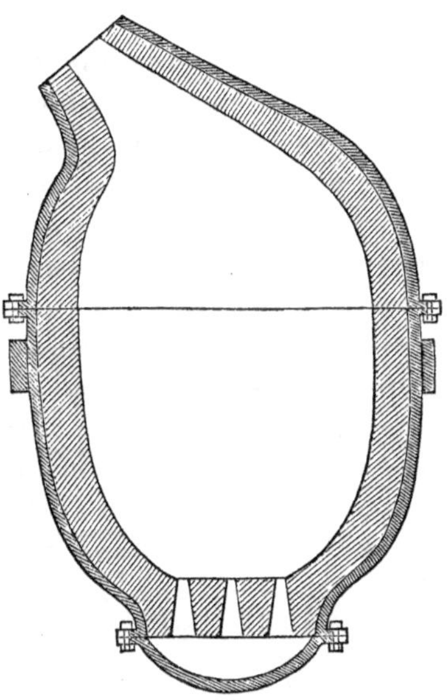

that part of the metal would be thrown out, if there were not plenty of room.

323. Everything being ready, the converter is placed in a horizontal position, and the charge of pig iron, previously smelted in a cupola or reverberatory furnace, is run into it by means of a trough lined with sand. The charge is then level with the tuyeres, and the blast is turned on before the converter is made to revolve to its vertical position, which is done slowly. After fifteen to twenty minutes of

blast, the converter is swung again to a horizontal position, in order to receive the additional charge of five to ten per cent. of spiegeleisen. Having again been made to assume the vertical position, after five minutes more of blast, the steel is completed and run into a large ladle supported by a crane. From this ladle the ingot moulds are filled.

The whole operation is one of the most impressive in iron metallurgy; torrents of sparks and flame escape from the mouth of the converter. The energy of the reaction diminishes as the decarburization progresses, but it is very difficult to ascertain exactly the proper time for emptying the converter of its contents. It is thought that spectral analysis will give the proper indication; at present, the guides are a certain duration of the blast for a given quantity of pig metal, and the appearance of the flame viewed with the naked eye or through different colored glasses superposed (blue and yellow) giving a dark neutral tint. Through these glasses, the flame appears white as long as the decarburization is going on, and turns red when all the carbon has been burnt off.

The oxidization of silicon takes place before that of carbon. The silica unites with oxide of iron, and forms a small quantity of slag. The pressure of the blast is about fifteen pounds to the square inch.

Taylor Process.

324. At the close of the year 1857, Mr. Taylor tried to employ atmospheric air for decarburizing pig metal and manufacturing steel, by submitting to a powerful blast, molten pig iron spread over a very large surface. We do not think this process more advantageous than that of Mr. Bessemer, but rather inferior; however, as it is very ingenious, we suppose its description will not be out of place here.

325. A semi-spherical kettle made of fire-brick or of metal lined with fire-brick, is inclosed in an arched space. The kettle is fixed at the end of a vertical shaft, and made to revolve horizontally with a rapidity of motion which may be varied at will. Above the kettle, and in the arch, an aperture is left for introducing the molten metal. This metal, falling into the rotating kettle, spreads itself against the sides, thus presenting a large surface to the action of the air. The decarburization is easy, and more or less rapid.

326. The rapidity of the refining depends on the velocity of the rotary motion of the apparatus, and on the quantity and pressure of the blast. By stopping the operation at a given time, a more or less decarburized metal, *i. e.*, steel, or ductile iron is produced. The process is, therefore, an ingenious modification of the Bessemer invention.

327. In this modification there is an improvement, *i. e.*, that by using the apparatus of Mr. Taylor, the manufacture of iron or steel is continuous, and, therefore, presents a great economy.

The pig metal becoming more and more decarburized, ascends the sides of the kettle, by centrifugal force, until it reaches the top edge over which it flows, and is projected to some distance against brick walls. Thence, the molten decarburized metal flows into a cavity, where it is collected for filling the moulds. The temperature increases during the operation, and the metal remains fluid all the time. As new quantities of raw metal are continually added, and the blast is steady, there is a continuous flow of liquid metal. This is not the case with the Bessemer process.

The apparatus is built in such a way that the vertical shaft and all the gearings are protected against the intense heat of the place; when necessary, cold water is run into the space left between the double walls.

Uchatius Process.

328. Mr. Franz Uchatius, captain in the Austrian army, manufactures steel from pig iron sufficiently decarburized with oxide of iron and some manganese.

329. Charcoal, pig iron, and powdered spathic ore are employed.

330. The first operation consists in *granulating the pig iron*, that is, reducing it to the size of shot. In this state, there is a greater surface to be acted upon by the oxygen of the ore, and the conversion is more rapid. The molten pig metal is made to fall into a tub of water[1] upon a broom which a workman is moving as near the surface as possible. The metal becomes very finely granulated.[2]

331. The theory of this process has been already explained at the beginning of this work. The excess of carbon in the pig iron is extracted by the oxygen of the ore. With the exact proportion of carbon the steel is hard, a little iron added to it makes it soft.

332. The granulated pig iron is put into a graphite crucible with some pulverized ore. If this is spathic, no manganese is needed, because it holds some already, otherwise, manganese is added.

[1] This process for granulating pig iron is well known. For a long time it has been employed for making spherical shot. In the year 1763 (July 29, No. 794), John and Charles Wood took out a patent in England for this purpose.

[2] The apparatus of M. de Rostaing for granulating metals would seem to be preferable. It consists of a metallic disk covered with refractory materials, and revolving with great rapidity in an inclosed space. By centrifugal action, the molten metal thrown upon it is projected in a very minute state against the walls of the room or cellar, and made to fall into water if desired.—*Trans.*

333. For manufacturing hard steel the proportions are:—

Granulated pig iron . . .	1000
Spathic ore	250
Manganese	15

334. An experiment made before a commission of French engineers has given the following results:—

Granulated pig iron . .	11.58 kilog.
Iron ore holding manganese	2.89 "

The operation lasted one hour and forty-five minutes, and the product was 12.40 kilogrammes of steel with a granular fracture, somewhat fibrous, and of a grayish color.

The proportions for middling hard or mild steel are the same; but to 100 parts of pig metal 12.5 parts of ductile iron are added.

The trial was made with the following quantities:—

Pig iron	12 kilog.
Ore	3 "
Pieces of iron	1 "

The operation lasted two hours and twenty-five minutes; and the steel weighed 14.85 kilogrammes. Its appearance was very much like the first sample, but of a lighter gray.

Soft steel was made with the following substances:—

Pig iron	10	kilog.
Ore	2.5	"
Iron	2	"

After two hours and eight minutes, the product was 12.70 kilogrammes of steel, more grained than the previous one, and bluish-gray.

335. Therefore, on an average, 13.32 kilogrammes of various kinds of steel have been produced in two hours and six minutes, or 100 kilogrammes in fifteen hours and forty-six minutes.

The expense in fuel has been, in weight, 2.30 of coke for one of cast steel, or 230 of coke, equivalent to 510 of pit coal, for 100 of steel.

VIII.

Damascus Steel.

336. *Damascus steel*, forged into thin blades, appears generally with veins and waving lines, well known, easily seen, and which indicate a metallic compound of excellent quality, very tenacious, hard, and difficult to break. In the Eastern countries this kind of steel is mostly applied to the manufacture of sabres and scimetars.

337. At Caboul Sir Alexander Burnes saw a scimetar valued at five thousand rupees ($2500), and two others estimated at fifteen hundred rupees each. The peculiar value of the former was due to the

great uniformity of its silky veins through its whole length. The value would have been a great deal less had the texture been intersected by transverse or angular lines. A sword of Persian manufacture, the lines of which were not continuous and parallel to the direction of the blade, was not highly priced. It belonged to Nadir-Shah. Another scimetar, from the Khorassan, did not present an elongated and undulating texture, but was dotted with small black spots. All these blades would vibrate like a bell when struck upon. It was asserted that they would improve by time.

338. For over half a century, efforts have been made to imitate the Damascus blades; often a very fine and variegated appearance has been obtained by piling (fagoting) and welding together steel and iron bars, or even different sorts of steel. All kinds of figures have been produced, waves, iridescent silky fibres, a twisted texture, mottled and dotted fibres, letters, inscriptions, leaves, flowers, &c., have been delineated with great perfection; but it has not been possible to produce a true Damascus steel, or blades having the same qualities as those manufactured in Persia, India, &c. From the ability exerted in Europe, and principally in France, in the effort to imitate the Damask steel, we may infer that the excellent quality of this product is not due so much to skill, as to the nature of the materials employed. Recent experiments have shown that when

the blades are cooled slowly, as by moving them in the air, the damaskeened appearance results from a large quantity of carbon. However, this is not so recent a discovery, because all the blades of Solingen, which are the nearest approach to Damascus steel, have been hardened in that way for centuries. This method evidently appears the best, when it is desired to preserve all the tenacity of the steel.

339. The damaskeened fibres will appear by washing the polished surface of the steel with diluted sulphuric or muriatic acid, which dissolves the soft parts of the steel, or those which hold less carbon. The steel is afterwards washed in pure water, dried, and covered with a film of oil or beeswax. We do not believe this is the method employed by the Orientals; it is more probable that, according to Aristotle, they bury their steel in the ground for a greater or less length of time.

340. Mr. Henri, of Bougival, has succeeded in manufacturing a damaskeened cast steel, entirely similar to the Eastern steel. We add here an extract of the *Bulletin de la Société d'Encouragement*, which gives an idea of this curious fabrication.

"A long series of experiments," the author says, "undertaken in view of elucidating the question, has demonstrated to me that the material of the Damascus steel is a cast steel, with more carbon than is found in our European steels, and into which, by proper cooling, have crystallized two distinct combinations of iron and carbon.

"This separation is the essential condition, because, if the molten metal is suddenly cooled, as in a small ingot mould, the damaskeened fibre appears under a magnifying glass only.

"The law discovered by Berzelius, by which a combination takes place between two bodies having some affinity, explains satisfactorily this characteristic property of Damascus steel, of showing a pattern on its polished surface, by the action of a very diluted acid.

"If the combination of bodies having some affinity take place only in definite proportions, all that is in excess of the proportion is not combined, but only mixed. Now iron and carbon form at least three distinct combinations: Steel, at one end of the series, contains only a very small quantity of carbon (one hundredth); on the contrary, in plumbago, there is twelve to fifteen times more carbon than iron; white and gray pig irons are intermediate.

"Let us suppose that in the manufacture of steel, the quantity of carbon is deficient; the quantity of steel will be in exact proportion to the quantity of carbon combined; the remainder will be iron mixed with it: therefore, by slow cooling, the molecules of steel being more fusible, will have a tendency to unite together and be separated from the iron. This alloy will produce a damaskeened pattern; but the pattern will be white, not very apparent, and the metal cannot become very hard, on account of being mixed with iron.

"If the proportion of carbon is precisely that necessary for converting all the iron into steel, there will be but one kind of combination; therefore, no separation of distinct compounds will take place during cooling. This, I presume, will indicate the proper proportion of carbon in the manufacture of the kind of steel best adapted to the working of metals.

"But, if the carbon is slightly in excess, all the iron will be converted first into steel; afterwards, the carbon remaining free in the crucible will combine in a new proportion with the steel already made. There will be two distinct compounds: pure steel and carburized steel or pig metal; these two compounds, intimately mixed at the beginning, will have a tendency to separate from each other by the molten mass standing undisturbed. Then a crystallization will take place, by which the molecules of the two compounds will aggregate, according to their affinity, and their specific gravity.

"If a blade, made of steel, thus prepared be dipped into acidulated water, a showy pattern will be revealed, in which the parts of pure steel will be black, and those of carburized steel will remain white, because the acidulated water has more difficulty in causing the carbon of the carburized steel to appear.

"The carbon, irregularly distributed in the metal, and forming two distinct combinations, is therefore the cause of the damaskeened pattern; and we can easily understand that the slower the cooling, the larger will be the damaskeened veins. On this account, it would probably be better not to melt too large quantities at once, or to modify somewhat the process. As agreeing with my opinion, I would name Tavernier, who, in his 'Journey in Persia,' has given some indications about the size of the lumps of steel which, in his time, were used for making Damascus blades.

"'The steel for damaskeening comes,' says he, 'from Golconda; it is found in the trade, in pieces as big as a one penny loaf of bread. They are cut in two, in order to ascertain if they are of the proper quality, and from each half, a sabre blade is made.'

"According to this narration it is apparent that the Golconda steel was in circular lumps like the Wootz, and that their weight was not over two or three kilogrammes.

"Tavernier adds that, 'if when hardening this steel, the European processes were followed, it would break like glass.' We must infer from this that it is very difficult to forge, an observation already made by Réaumur.

"This savant, having received from Cairo some samples of Indian steel, could not find anybody in Paris able to forge them. On that subject, he says that the fault is in our workmen, because the Orientals are able to work that kind of steel.

"As carbon is the essential part, not only in the

formation of the pattern, but also in the intrinsic qualities of the steel, it is to be supposed that Messrs. Stodard and Faraday have been mistaken in their researches, the same as I have been for a long time, when they attribute to metallic alloys effects more particularly due to an excess of carbon.

"I am very far from contesting the presence of metallic alloys in the Oriental sabres, although in the few samples I have been able to analyze, I never found silver, gold, palladium, nor rhodium; nevertheless it seems to me very probable that such combinations have been attempted. Indeed the same people who had succeeded in hardening copper by alloys, must have tried similar processes with iron.

"Following that idea, I have formed various metallic alloys, some of them giving satisfactory results. One of the sabre blades I have exhibited, contains one-half of one per cent. of platinum and a greater proportion of carbon than is to be found in ordinary steels. It is to that excess of carbon that the pattern is mostly due. Some excellent razors have been made with the same alloy."

341. Experiments made by smelting together pig-iron, carbon, and alumina, produced a highly aluminous steel which was welded and drawn with a steel of cementation. From this mixture a steel was obtained very much like the Wootz, and producing immediately a damaskeened pattern. However, skil-

ful metallurgists assert that aluminum is not indispensable for the manufacture of Damascus steel.

IX.

Intermixed Metals (Etoffes).

342. The *intermixed metallic tissues* are alloys of steel with one or several metals, producing a metal whose fibres present different patterns, and are elongated, interlaced, or zigzag.[1]

343. The experiments of Messrs. Stodard and Faraday, and those of Guyton-Morveau, have demonstrated that steel may be alloyed intimately with silver, gold, platinum, rhodium, nickel, and copper. On such authority new alloys have been tried everywhere, to which, by the way, too great merits have been attributed. The French manufacturers, according to foreigners, are those who have spent the most money and time in such experiments.[2]

344. An alloy which deserves great notice is that of silver with steel. The former of these metals, as is well known, has a tendency to separate from steel in the form of thread and drops. Therefore, when the alloy is heated and kept fluid for a certain length of time, it seems perfectly homogeneous and com-

[1] Wire twist, stub twist, stub Damascus, &c., for guns, are metallic tissues of iron and steel.—*Trans.*

[2] Frederick Overman, Philadelphia.

pact; but by cooling and solidifying, the silver seems to ooze through the metallic texture, and appears in small separate drops. When, in forging, the heat is slow and moderate, instead of drops, filaments will appear as slender and elongated as those of the capillary silver in certain kinds of silver ores.

345. Therefore, silver does not alloy chemically with steel and iron; 1 part of silver and 100, 200, and 400 parts of steel do not produce an intimate union, and the silver is constantly separated in the form of filaments. For an intimate alloy, 500 parts of steel and 1 of silver are necessary.

This alloy has a very fine appearance; it is so hard that, in this respect, it ranks above the best cast steel, even the Wootz. It does not crack by hardening, nor by hammering, and produces, by forging, tools and instruments perfect in quality and excellent for use.

346. The English metallurgists prefer the rhodium steel, it being harder. This alloy contains 1 to 3 per cent. of rhodium, and requires to be tempered at a higher temperature than is necessary with cast steel. Such is the tenacity of rhodium steel, that cutting instruments made of it will bear a tempering of 30° Fahrenheit above that given to the best Wootz. Its damaskeened patterns are very fine. An alloy of steel with 1.5 per cent. of rhodium has a specific gravity of 7.795.

347. Steel alloyed with platinum is not so hard as the silver steel alloy, but has a greater tenacity. The two metals appear to unite in every proportion, and when the fusion has been complete, no separation takes place, as is the case with silver and steel. The metallic compound is perfectly homogeneous.

348. Equal parts of steel and platinum produce excellent mirrors, which will polish well, and will not tarnish. The specific gravity of the alloy is 9.862 before forging.

With only 10 per cent. of platinum the alloy will not tarnish, and will receive a polish fine enough for mirrors.

For cutting instruments, the alloy which appears the most proper, contains from 1 to 3 per cent. of platinum.

The characteristic property of the alloys of steel with platinum is their resistance to oxidation.

349. Chromium and steel give an alloy with some valuable properties in certain cases. It is rather difficult to produce their intimate union; nevertheless, with care and some precautions, this can be done. The process employed by Mr. Berthier, who was the first to make useful experiments on chromium steel, is as follows:—

He mixed 10 parts of the natural chrome iron ore with 6 parts of iron scales, and 10 parts of glass free from metallic substances. The whole was smelted in a brasqued crucible in a wind furnace. The result was a lump weighing 7 parts.

This alloy was then combined with steel in the proportion of 1 to 1.5 per cent. of chromium. The chromium steel thus manufactured is excellent. It can be forged, and presents a fine damaskeened pattern, if, after polishing, it is treated with diluted sulphuric acid. The veins are of a bright silver color, very much like those of silver steel, but very probably they are pure chromium.

PART THIRD.

WORKING OF STEEL.

---

350. Drawing or tilting an impure and heterogeneous steel, when cast steel is not at hand; welding together several pieces of steel, or steel to iron by way of economy; or annealing a steel too harsh or too hard; hardening it when too soft; giving to it by fire, and after hardening, the degree of temper proper to the various duties it has to perform; compressing its texture and at the same time giving it a regular and straight shape; such are the six manipulations to which steel is generally subjected.

We will make six chapters of these manipulations under the titles of *Refining by Drawing or Tilting*, *Welding*, *Annealing*, *Hardening*, *Tempering*, *Hammer Hardening*.

I.

Refining by Drawing or Tilting.

351. The processes for manufacturing natural steel are so incomplete and uncertain, that very rarely a

homogeneous, tenacious, and elastic steel is thus obtained. Where it is not possible or not required to perfect its homogeneousness by fusion, it must, however, before being delivered to the trade, be drawn the same as iron, in order to give more regularity and a more uniform composition to its texture. This drawing is sometimes termed *refining*.

The number of heats given to steel during drawing, depends on the quality of the crude steel; the more homogeneous it is the less drawing it requires. An exposure to the fire, too often repeated, will burn the carbon, and it may happen that the nature of the metal will be entirely changed.

352. The blooms are drawn into slabs or flat bars 0.55 to 0.66 metre long, and 0.40 to 0.50 metre wide, which are plunged into cold water when red hot. Afterwards they are piled up, taking care to match them well. The practice of the workman makes this easy by judging the fracture of the slabs. The outside slabs are of one piece, but the inside ones may have various dimensions, and may contain broken pieces.

The bundles or fagots are put into a reheating furnace, heated to the welding point, and well sprinkled with clay. This is done to cover them with a layer of slag which protects them against the decarburizing action of air. The bundle is then carried to the hammer or the rollers, where it is drawn into a square slab 0.04 to 0.045 metre thick, which is afterwards cut in two, doubled, welded, drawn again, &c. The same operation takes place three or four times.

353. The furnaces employed are similar to the *reheating furnaces* for puddled iron. Sometimes they are forge fires covered with a depressed arch. In the former, pit coal is burned; in the latter, charcoal or very pure coke.

354. The experience of the workman who makes the bundles has a great deal to do with the improvement of the steel; if he is skilful, he may remedy the bad quality of steel by a judicious choice of slabs, when, however, the bad quality does not come from the raw metal itself.

Hammered or *tilted steel* is considered better than rolled. *Shear steel* is a steel of cementation which has been piled, welded, drawn, doubled, &c., one or more times, according to its quality.

II.

Welding.

355. The relatively low temperature at which steel loses its carbon and becomes less valuable, is the great difficulty in welding steel to iron. If the workman heats the former metal too much, the operation fails, and it is customary to charge the result of unskilfulness to the bad quality of the steel. Those who succeed by practice and knowledge of

the materials they employ, pretend to possess a secret, an herb, a salt, or something else which they have discovered or inherited. This is the secret:—

356. The iron must receive a glaring welding heat, while the steel is heated a great deal less. We must bear in mind, that when the two bars are brought into contact, their two temperatures will become equipoised, and if then the temperature of the steel is too high, it will lose its quality. The proper degree of temperature for steel is cherry red; above this it will entirely lose its tenacity, and crumble to pieces under the hammer. On the contrary, iron is welded at a glaring white heat; and it may yet support a small increase of temperature before melting. There is a proper time for setting the piece of steel, which requires all the attention of the workman.[1]

357. The respective masses of iron and steel must also be considered in the operation. If the volume of steel is small compared with that of the iron to

[1] Greater difficulties are encountered when the chemical composition, *i. e.*, proportions of carbon, of steel and iron are too remote. For instance, a very hard cast steel (highly carburized) will be difficult to weld with a very soft, fibrous iron (scarcely carburized). Less carburized steels, such as shear steel, mild steel, natural steel, will weld readily with iron. But with very hard or harsh steel a certain kind of iron, called steely iron, will be found useful, having a composition intermediate between iron and steel.—*Trans.*

which it is to be welded, the former metal might be heated too much by absorbing part of the heat of the iron. Therefore, when a small piece of steel is to be welded to a large piece of iron, a careful workman will heat thoroughly only the part where the steel is to be applied, in order that the equilibrium of temperature may be distributed in the whole mass, and not in the steel alone. If the piece of steel is larger than the iron, it should be heated as much as practicable.

358. Formerly, and with good results, borax was used for welding steel to iron, this flux lowering the point of fusion of steel. At present, it is employed only when welding steel to steel.

III.

Annealing.

359. We must not confound the *tempering* given to steel after hardening, with the *annealing* given to the unhardened metal in order to make it softer under the file.

360. Tempering after hardening, as will be explained (384), diminishes the brittleness and hardness of steel, at the same time giving it body and some elasticity. This result once obtained, the steel is to remain in the state acquired by the tempering. By annealing in the forge, steel is only prepared for

the hammer or the file, and it may, and often must be hardened afterwards.

361. Annealing steel is useful; but the fire must not be pushed to the utmost, as is commonly done. It seldom happens that steel is pure, *i. e.*, contains exactly the proportion of carbon which constitutes the definite alloy (153, 167); generally there is an excess of carbon. As this excess of carbon has less affinity for the metal than the carbon of the definite alloy, it may be easily expelled at a moderate heat, where the oxygen of the air is the decarburizing agent. But if the temperature is too high, part of the carbon of the alloy is burned out, the iron itself is oxidized and covered with scales, and the steel becomes deteriorated, loses its nature, softens too much, and diminishes considerably in weight.

Some manufacturers well conversant with the working of steel, have adopted the following rational method for annealing:—

362. They heat the steel to *dark red*, called by some *blood red heat,* and afterwards plunge it quickly into charcoal dust, where it cools. Others dip it into water.

Annealing in water is a hardening at a low temperature, which, as will be seen in our chapter on hardening, produces a steel soft, and easily filed.

363. Annealing in charcoal dust is a sort of

cementation which, on the contrary, will increase the hardness of steel, by taking off part of its homogeneousness. These defects may be corrected by hammering, but it is preferable to avoid them.

IV.

Hardening.

364. When steel is heated, it expands, and its constituent principles, iron and carbon, take a different grouping from that which they had when cold; the carbon in the amorphic state has a tendency to crystallize, thus becoming hard; and the electricity which is developed in iron up to a cherry-red heat, produces in the metal a crystallized texture which is favored by a previous hammer hardening. All parts are in a state of extreme tension.

365. If, after being heated to a cherry-red, steel is allowed to cool slowly, the texture does not return entirely to its primitive state; it is less crystalline, and would be fibrous, were it not for the presence of carbon. The result is a partial softening and more ductility, due to a less quantity of carbon on the surface, and to a tendency of the iron to have its fibres elongated.

366. If, instead of allowing steel to soften by a slow and progressive cooling, it is plunged into a cold liquid, such as water, at exactly that moment when all the molecules are strongly extended, and in a

sort of confusion due to the unequal dilatation of their elements, all movement ceases. The molecules of the two elements of steel remain in the position they had acquired, the texture remains granular, and the steel becomes hard.

367. Such is the phenomenon of *hardening*, many times explained in a more or less satisfactory way.

268. Hardening is the operation by which steel is rendered hard. At least, this practical definition cannot be criticized.

369. It is evident that the metal, having extended, previous to its immersion in cold water, and having retained its texture as it was before the sudden cooling, the hardened steel has lost its primitive specific gravity, and has therefore increased in volume.

370. Red-hot steel dipped into tepid water, does not gain much hardness; the mass is not suddenly cooled, and the texture is somewhat changed. Oils, tallow, and most fatty matters, which become hot, and are even vaporized by contact with the burning metal, produce only a feeble hardening. Generally, the colder the liquid, the greater is the hardening, because its action upon the texture of the steel is more rapid.[1]

[1] Also, certain metals or liquids which have a greater conductibility for heat than water, will produce hardening, by sudden absorption of the heat of steel. Such are quicksilver, and some saline or acid solutions.—*Trans.*

371. Therefore, we may consider as a settled fact, that *the hardness of steel is in proportion to the difference of the extreme temperatures of the heated steel, and of the liquid into which it has been immersed.*

372. However, this theory has some limits: hardened at too high a temperature, that is, when the dilatation has separated the molecules of steel too much, the grains or confused crystals thus produced are no longer able to retain the elasticity in the metal. These grains are larger than usual, rough and bright; and the steel is brittle, dry, and easily crumbled to pieces under the hammer.

373. A *cherry-red heat* appears generally to be the most proper for hardening: at such a temperature, cast steel and other good steels acquire the maximum of hardness, and present a fine granular appearance. With a dark red heat, hardening does not produce much effect, the steel remains soft; with common kinds of steel, the grains are irregular and intermixed with particles of iron. When the temperature has been raised too high, steel is injured, and loses a portion of its tenacity and hardness.

374. Notwithstanding what has been said, and the so called experience of some practical metallurgists, pure water is the best liquid for hardening steel. It is a mistake to believe, with the ancients, that certain waters are more adapted to this operation

than others. The only difference lies in their temperature. A workman of Caen, Mr. Damesme, who has published a diffuse work on steel, has tried the hardening of steel in the juices of vegetables, and has ascertained that there is comparatively no advantage over hardening in water. Mercury has no other property than that of being cold, and of producing a hardness which can be obtained with water at the same temperature. Tallow and oils, where carbon is one of the constituent elements, produce an imperfect hardening, but prevent a loss of carbon. When by overheating, steel has been burned and decarburized, the oils and fatty matters are useful, because they give back to the steel a part of the carbon lost in the fire. Some acids, such as sulphuric, are justly considered as imparting more hardness to steel, by dissolving a film of iron from the surface and exposing the carbon. As for urine, alcohol, brandy, and a thousand other liquids extolled by ignorant workmen, they are not worth as much as water, which has the advantage of being abundant everywhere, cheap, and adapted to all changes of temperature.

375. Steel should be hardened to the point corresponding to its nature and its use. Indeed, it is possible to correct the quality, either by increasing the hardness by a very cold dipping liquid, or by producing more elasticity when tempering; but these corrections are left too much to the judgment of the

workman to be considered efficacious. For instance, in fine cutlery, and principally in the manufacture of surgical instruments, every instrument must have its peculiar hardness and tenacity. Very few men always succeed in the operation, which, generally, is left to chance.

Hammers, cold chisels for iron, drills, engraving tools, require a strong hardening, a great hardness; sabres, razors, straw-cutters, &c., do not require to be dipped into very cold water; table-knives, scissors, and springs, require less hardness.

376. We readily understand, that if the temperature the most proper for the degree of hardness and tenacity of the instrument were known, it would be sufficient to raise the instrument to that temperature, and to immerse it afterwards in water.

377. Some workmen heat the steel which is to be hardened, much above a cherry-redness, allow it to cool slowly in the air, and wait till it has taken a certain color, previous to plunging it into water. This is a very bad practice, because, by an excess of heat, there is a loss of carbon, and an alteration of the steel, which has then large grains, and is without tenacity at the edges.

378. In order to graduate the heat, and to bring the instruments to various and distinct temperatures, D. Hartley, in 1789, thought of using a pyrometer,

when hardening. This process, very good, indeed, was difficult in practice. Sir Parkes was more successful, by determining in advance the various points of fusion and of perfect liquidity of certain metallic alloys. These temperatures being known, steel is plunged into the molten alloy, the same as into a forge-fire, and when thoroughly heated, is dipped into cold water.

379. Although this method has not been generally employed, for the sake of its ingenuity, we will take from the compositions of Sir Parkes, those which most nearly correspond with the various colors and temperatures necessary for certain instruments.

The temperatures are in degrees centigrade:—

Lead.	Tin.	Temperature of fusion.
7 parts.	4 parts.	213.40°
7½ "	4 "	221.11°
8 "	4 "	225.50°
8½ "	4 "	232.22°
10 "	4 "	240.90°
14 "	4 "	251.90°
19 "	4 "	262.35°
30 "	4 "	273.90°
48 "	4 "	284.90°
50 "	4 "	289.20°

Linseed oil boils at 312.40°.

Lead melts at 319°.[1]

[1] The metallic baths above named are certainly not for heating steel previous to hardening, but for tempering steel already hard-

V.

Tempering.

380. Hardened steel is generally harsh and brittle; so is chilled iron, probably for the same cause.

ened. For hardening, steel should be at a cherry-red heat, or much above the temperature of any of these metallic baths, which, besides, do not remain at the temperature first indicated, whether by oxidation or by some molecular changes. This has been observed many times in the "safety metallic plugs or plates" attached to steam boilers.

Pure lead alone has been employed for heating certain delicate pieces of steel previous to hardening; but for that, the temperature of the lead bath should be raised above the melting point, which will be best seen in a dark room.

Hardened steel expands, as is well known by those who case harden finished pieces of machinery which fit close. Experiments by Captain Caron have shown that hammered bar steel will not extend in length, but will extend in the other directions. Rolled sheet steel and steel wire extend in all directions.

We have seen that the molecules of hardened steel are in a state of extreme tension, often producing breaks in pieces of large dimension. To avoid this trouble, it has been proposed, and successfully tried, to compress or condense rapidly the heated piece by the hammer or otherwise, before plunging it into cold water.

It seems to us that in regular and continuous operations, the ordinary forge-fire could be advantageously replaced by a gas fire, heating the piece directly, or in a muffle. For small objects and small workshops, illuminating gas might be employed. In larger establishments, some gas generating furnace on the principle of Siemen's gas furnace, could be devised. In all such apparatus, where the flow or the production of gas may be regulated at will, and when regulated, the heat is constant, there would be much more certainty and facility in ascertaining the proper temperature for hardening and tempering. There would be also economy.—*Trans.*

381. The natural structure of iron is crystalline; it becomes fibrous only by artificial means, but returns to its former texture under certain conditions. Percussion at a cold temperature, vibrations, sudden cooling, even a protracted rest, cause the action of magnetism to which the confused crystallization of the metal is due. The natural state of carbon is a crystalline texture; it is also essentially electro-negative. Doctor Ure has demonstrated that negative electricity produces immediately a crystalline arrangement: hence, we may infer that carbides of iron cannot have naturally a fibrous texture, and this explains how steel is always granular.

382. Heat is also one powerful agent in changing the structure of iron; and if all pure metals, when melted, acquire a crystalline texture by cooling, they will become fibrous by being hammered when hot, and allowed to cool slowly after another heating.

383. In hammered or tilted steel, the iron retains its property of becoming elongated under the action of the hammer; but this property is counteracted by carbon, whose tendency is to remain crystalline. However, if after having rendered it crystalline by hardening, the metal is reheated and allowed to cool slowly, it has a tendency to become fibrous and to acquire body and tenacity.

384. This improvement in the tissue of steel has been made use of by practical metallurgists, who

after hardening, the immediate result of which is brittleness, reheat the metal and allow it to cool slowly, in order to give it body and fineness.

This operation is termed *tempering.* The result of tempering is to remove part of the brittleness, and especially to graduate the hardness of the steel to the degree wanted, and to give it some elasticity.

385. If, after a strong hardening, which will be the type of extreme hardness, steel is heated again to redness, it loses all the hardness it had gained, becomes soft, and will be rendered hard again only by a new hardening.

Between these two extremes: hardness and softness, there are several degrees which are as many shades of the qualities adapted to certain uses.

386. These degrees are made apparent by the color of the metal when reheated, and take place in the following order:—

1. Being put upon burning fuel, the steel gradually heated becomes tarnished, yellow, and *straw-yellow.*

2. The heat increasing, the color deepens, and reaches a gold yellow, *full yellow.*

3. Afterwards, the steel takes several shades, rapidly following and blending with each other; they are purple, pigeon's throat, copper, *brown purple.*

4. These shades become deeper until they become *violet.*

5. Afterwards, they pass rapidly to indigo blue, *full blue, dark blue.*

6. This color becomes weaker, and gives a *sky blue* more or less pure.

7. The blue takes a greenish tint and produces shades which are gray and *sea-green.*

8. At last, the steel *reddens*, and will no longer give distinct colors.

The shades of these eight colors, which are called *tempering colors*, are perfectly distinct, very apparent, and easy to recognize; but they take place only after hardening and on clean steel. The metal which has not been hardened, will not show these colors so plainly; the shades are mingled, blended, and less in number.

387. The colors, during the tempering, are a sure guide for the workman, of the degree of hardness or tenacity he desires to obtain. Dark blue indicates a great tenacity, straw-yellow produces a greater hardness, and is the tempering shade for razors. Bistouries, lancets, penknives, erasing knives, some scissors, and generally blades requiring body, are reheated to full yellow. The strong blades for table knives, and gardening tools, are tempered to a brown or purple brown. Purple is the proper color for large shears. Violet and dark blue are for springs; with a violet color, the spring will be very elastic but brittle, a blue shade will make it very resisting. It is very difficult to break a spring reheated to the color of water; but its elasticity is a great deal lessened.

The temperatures (centigrade) corresponding to these colors, and best adapted to the tempering of various instruments are seen in the following table:—[1]

Lancets	210°—215°
Other surgical instruments	220
Razors	225
Penknives, erasers	230—235
Scalpels, cold chisels for iron . . .	240
Shears, sheep shears, gardening tools .	250
Hatchets, axes, plane irons, pocket-knives	260—265
Table knives, large scissors . . .	270—275
Swords, watch springs	285
Large springs, daggers, augers . . .	290
Saws, some springs	310—315
Various other instruments requiring less hardening	320

388. The hardened instruments are reheated in or upon a live fire, easily regulated, and without the help of bellows as far as practicable. An intelligent workman will cease blowing as soon as he perceives that the metal begins to change its color.

The proper shade must come by itself without increasing the fire, and must be regular all over, before the piece is plunged into cold water. Sometimes, this last dipping is omitted.

The small pieces, such as penknives, erasing knives, &c., rest upon a wire cloth put into the middle of the fire; when they have reached the proper color, they are cooled in water.

[1] See also paragraphs 378, 379.—*Trans.*

A lancet requires a special tempering: the shank must be blue; from there the color will be first purple, next brown, and at the point, full yellow. These various shades upon one blade are a necessity, on account of the degree of hardness and tenacity required by this instrument. Full yellow will produce the proper sharpness, but would not be suitable to the rest of the blade, which, instead of hardness, must have tenacity and elasticity.

A good workman, willing to give the greatest perfection to an instrument, will be very careful when tempering it, in order to obtain the various shades which are necessary. A knife, for instance, must be brown purple at the cutting edge, purple in the middle, and sea green at the back, to unite the hardness of the cutting edge with a certain amount of resistance which will prevent its breaking under a strain.

This is obtained by using certain precautions, and above all, by not going beyond the proper degree, because it is very difficult to retrace the steps. If the fire is too strong or irregular, part of the edge may be purple brown, while the other is only straw-yellow; then, by pinching the blade between red-hot tongues, at the place which should be more heated, the temperature rises rapidly, and the instrument is brought up to the proper tempering point.

Certain scraping and burnishing tools, and steels for sharpening, do not require any tempering, because they cannot be too hard.

It happens, though rarely, that steel bars which have been hardened and left for some time in store rooms, will break with a noise, and will project to a distance, pieces of steel from the corners. This phenomenon does not take place with small pieces, such as smooth or even bastard files, but will happen with large rubber files, mostly those of cemented steel.

By hardening too quickly, the same effect is sometimes produced; the workman receives a shock in his arm at the moment of dipping: part of the piece breaks off with a noise, or the steel splits along its length.

VI.

Hammer Hardening.

389. *Hammer hardening* or *cold hammering* is injurious to ductile iron, which it renders granular and cold short.

390. On the contrary, hammer hardening improves the carburized metal, by increasing its tenacity and hardness, and by producing elasticity and toughness.

391. Hammer hardening a piece of steel should not be done too rapidly, otherwise the heat produced would diminish the hardness.

392. It often occurs that a thin piece of steel will become distorted or curved in the water, when hardened, and will retain that shape. The cause of this distortion is not well known; but, if it is noticed that it is the more frequent as the blacksmith is the less skilful, we would attribute it to an irregular percussion, a bad hammering, and a wrong mode of working in the forge.

393. Without any doubt, a bad hammer hardening is one of the most frequent causes of distortion. An irregular hammering produces unequal density and hardness in various parts of the piece of steel. By subsequent heating this piece extends unequally; and, when it is about to be plunged into water, it is already disposed to become twisted and distorted.

We shall indicate in the chapter on files, how twisted and distorted steel may be straightened.

PART FOURTH.

PROPERTIES OF STEEL AND ITS USES.

---

394. BY giving here a general knowledge of the properties of steel, of its qualities, and of some of its uses, we do not pretend to describe all the uses to which this metal is, or may be applied. We shall restrict ourselves to some facts which may guide the workman in choosing and discerning the good from bad steel, in appreciating its resistance, and in avoiding certain accidents. Among the uses of steel, we shall dwell more fully on the important manufacture of files.

I.

Characteristics of Steel.

395. Steel may be distinguished from iron by means of diluted nitric acid. One drop of it upon steel will produce a stain of a dark gray color, while upon ductile iron the stain is greenish.

396. The specific gravity of steel is too near that of iron (7.788) to be a distinct characteristic. Each

sort of steel has also its peculiar specific gravity, as may be seen by the following numbers:—

Natural steel	7.500
Huntsman's tilted steel . . .	7.900
Œregrund steel	7.313
Cast steel	7.800
Wootz (raw steel)	7.181
" forged	7.647
" cast	7.200

397. The coarsest kind of steel, the grains of which are the most marked, has a great tendency to take a fine granular texture by hammering. Let us take a steel of cementation with coarse grains, of a silver gray color, breaking without being previously notched; if we flatten and smooth one end of the bar by a light hammering, and afterwards cut this part, we will be astonished to find it has acquired body, a fine granular and homogeneous texture, and a sky blue color or light gray.

398. Puddled steel is perfectly crystalline; its grains are even finer than those of cemented steel. It is homogeneous in its fracture, very sonorous, and will take all the degrees of hardening and tempering; it possesses the same elasticity as ordinary steel, and may be transformed immediately into tools and large pieces of hardware, for which it seems exceedingly well adapted; but it has not yet been used for fine cutlery.

399. Steel has sometimes such a near approach to iron that it has a tendency to take a fibrous texture. With puddled steel particularly, a bar may be bent when cold, and without breaking. If it is bent in the opposite direction it will show a fibrous texture. A piece of steel plate, notched with a chisel and broken, has a similar appearance at the fracture; but if it is forged, hardened, and tempered, the natural crystalline texture will reappear.

400. It might be useful to be able to compare the hardness of various commercial steels, or that of hardened instruments. The hardness being the resistance of the metal to scratching, polishing, or cutting by a harder substance, this measure will be the strain necessary for cutting and scratching it.

401. The softest hardened steel may be scratched by glass, the hardest steel can be scratched only by the diamond. Between these two extremes we may adopt four degrees of hardness, and ascertain them by certain substances, so chosen that the preceding one will be scratched by the following one. By trying to scratch with these types of hardness six different sorts of steel, it is easy to find out the proper number in the series, and to appreciate the comparative hardness of steel with quite a mathematical accuracy.

402. These are the substances proposed in their order, or scale of hardness:—

1. Glass.
2. Feldspar adularia (subtranslucent).
3. Quartz.
4. Yellow topaz from Brazil.
5. Corundum.
6. Diamond.

403. There are some signs which should not be overlooked by a workman when he chooses the bars of steel. The sound is one of the principal indications; when silver-like, vibrating, and protracted, it shows a good quality; but when dull, hollow, and rapidly lost, it points to a want of homogeneousness. The fracture must present a fine round grain, easily seen, and with a lustre rather dull than shining. The finest steels have a white grain, difficult to be seen; but they have less body, and will not endure many reheatings. Where the outer surface is smooth and well hammered, the texture is generally uniform and homogeneous, although this outside indication is not sufficient to prove a superior quality of steel.

404. It is a mistake to believe that the finest steel is granular only, without fibres: its confused crystallization does not prevent the tissue from having a certain direction, generally in the way the drawing has been effected. Indeed, the cracks which take

place in steel assume this direction. Therefore, it is a good precaution to consider this direction of the fibre when forging, not only when using steel of cementation, where a fibrous direction is generally perceived; but also, with a cast steel the grains of which can scarcely be perceived.

405. A steel which welds easily is, with reason, preferred. This quality will be ascertained by heating two pieces of the same bar and welding rapidly their "tapering scarfs." If, after forging, the faces are neat and without cracks, the steel is of good quality.

Of course, we suppose that the blacksmith is skilful; otherwise, it is very easy for an awkward one to burn the steel, especially when it is finely granular, and to render it impossible to weld or *harsh*. When the metal has been properly hammered, and after it has been heated to the welding point, if cracks again appear, the steel is harsh and hard to forge. This defect is seen in steels which, having been doubled, break at the bend.

406. Soft or *mild steel,* after a welding heat, may be forged flat without further heating. By cooling under the hammer, it will not crack at the edges. If, after having been forged in this way, reheated to a dark red (blood-red heat), and plunged into water, it will bear cold hammering, bending in various

directions without cracking, its quality is perfect, and corresponds to that of a fine mild steel.

407. Certain sorts of steel swell by the heat, boil and project sparks in the forge fire, producing at the same time a kind of hissing noise. By attempting to forge them, they crack under the hammer, and are soon entirely broken.

408. The hardness of the steel is also a quality which should be ascertained, with reference to its special use. It is possible, by hardening, to ascertain if this hardness is uniform, and, of course, if the whole tissue is homogeneous.

For this purpose a bar of steel is put into the fire, and hammered at one end for a length of 0.15 to 0.18 metre. This portion is afterwards divided into six parts by notches made with a cold chisel, and penetrating one-fourth of the thickness of the steel. The bar is again put into the fire, heated gradually, in order to give a saffron color at the flattened end and a lesser temperature to the remainder, and then plunged into cold water, where the steel acquires various degrees of hardness. These degrees may then be ascertained, either with a file, which will give an approximate comparison, or with the substances we have indicated in the paragraph 402.

409. After the trial of hardness, the same bar may be used for the trial of the texture. This is

done by putting the bar of steel into a vice, and striking off with a hammer each of the notched parts. These being put along side of each other in the order they occupied on the bar, their fracture may be compared. It will be noticed that the grain will be the finer as the broken piece has been less heated. The lower the temperature of hardening, the finer is the grain.

410. The quality of the steel, as regards welding, hardness, and fineness of the grain, may be ascertained by the following process:—

A small bar of steel, not very thick, is welded upon a bar of iron having the same width but half the thickness. This iron is afterwards cut along its entire length down to the steel.

The whole is then heated and hardened. By keeping the welded metals on the edge of an anvil, the iron up, and striking with a hammer, the steel is broken in the direction of the cut. An examination of the fracture will show, at a glance, the texture, the grain, and the quality of the steel.

411. We cannot afford here to ignore a phenomenon which, for a long time, has been considered either an error or a falsehood, which has tried the sagacity of practical metallurgists, and yet remains unexplained.

412. According to Diodorus and Plutarch, the Celtiberians buried their steel in the ground, leaving

it there until it was strongly oxidized. From this steel, they forged swords and other weapons, the hardening of which was so perfect, that helmets and shields could be cut with them.

413. The ancient cutlers of Sheffield, so celebrated for their products and the good quality of their instruments, were also in the habit of burying in the mud of some stream or pool their bundles of steel, which remained there for several weeks. They affirmed that the metal, which had lost much in weight, had gained considerably in quality. When scissors manufacturers suspect that their steel is brittle, they bury it in their cellars; and they assert that it becomes more ductile and tenacious.

414. In the new edition of the *Manuel du Maître de Forges*, Paris, 1858, my father has recorded what happened to Mr. Weiss, a cutler of London, who having converted into steel the old iron of the piles of the city bridge, buried there for over a century, asserted that he had never since been able to find steel so fine and of such excellent quality.

415. The trials made on the tensile strength of steel, that is, the resistance to tension in the direction of the length of the bars, give results varying greatly according as the metal is hardened or not.

A strongly hardened steel has less tenacity than a steel which has not been hardened; but, if the bar has been properly tempered, its tenacity is a great deal superior to what could be expected, and is not in proportion to the tensile strength of steel hardened or not.

Square bars of steel of 0.0062 metre a side, submitted to traction, have given the following numbers, which are the breaking weights per square millimetre:—

	Kilog.
Ordinary steel not hardened	73.96
Middling " " "	84.92
Very good " " "	81.49
" hardened and not tempered .	76.70
" hardened and slightly tempered	102.72
" hardened and more tempered .	92.45

These experiments are due to Muschenbroeck. Others made by Rennie with square bars of 0.00635 metre a side, have given, always by square millimetre of section:—

Cast-steel, tilted . . .	93.79 kilog.
Cemented steel, tilted . .	92.93 "
Forged steel, drawn . .	89.07 "

Very possibly, the bars had not been hardened, although Rennie does not mention the fact.

416. Tires for railroad wheels require a hard, compact, and uniform material. It must be hard, because there is greater resistance to wear; compact, because the wheel revolves with less friction and more facility; uniform in its texture, because the wear should be regular.

417. These three conditions are not fulfilled with iron tires: the metal is not very hard, and softens by heat; it is not compact and homogeneous, presents parts which are soft and irregularly welded, and therefore, the wear is not uniform.

Iron tires worn about 0.0022 to 0.0032 metre must be turned again; and on account of the irregularity of the wear, the tool must remove a thickness of metal of about 0.0064 metre (6.4 millimetres).

418. Those of puddled steel are returned to the lathe, only when the wear is 0.0027 metre; the chips taken off, are 0.0055 metre thick.

Cast steel tires are worn so regularly, that when repaired, only a thickness of 0.0038 metre is removed.

Previous to repairing, *i. e.*, when the wear is respectively of 0.0022, 0.0027 metre, the tire must have run:—

Iron tire . . .	18,831 kilometres.
Puddled steel tire .	40,487 "

419. Considering the comparative duration of these different tires, the thickness they require in proportion to the material of which they are made, and the number of times they may be put upon the lathe, we find:—

420. An iron tire is 0.050 to 0.0523 metre thick, and may be turned five times and worn six times.

421. A puddled steel tire is 0.039 to 0.040 metre thick, and may be turned four times and worn six times.

422. A cast steel tire of 0.0283 to 0.030 metre thickness, may be also returned to the lathe four times and worn five times.

423. All things considered, we find that a tire, before it is put completely out of use, will have made runs:—

Iron tire . . .	141,240 kilometres.
Puddled steel tire .	202,435 "
Cast steel tire . .	706,200 "

424. The great elasticity of steel has been made use of for manufacturing horseshoes.

The horny matter of the horse's foot, which is in immediate contact with the ground, has the property of contracting when the weight of the animal is upon it, and of expanding when the foot is raised. The ordinary iron shoes do not allow this elasticity, and the result is an incessant friction and the production of callosities, which very much impede, if not disable the animal. The ossification of the cartilaginous parts is the cause of great suffering, and somotimes of premature old age.

An English veterinary surgeon, Mr. Bracy Clark, reflecting that of all domestic animals, the horse suffers the most from diseases of the feet, and is also the only one wearing a shoe without expansion, suggested, therefore, the manufacturing of steel horseshoes, called *the steel tablet expansion shoe*, and made of several plates of spring steel of proper thickness. After numerous trials and improvements, the inventor succeeded in producing horseshoes answering the purpose very well.

425. England, notwithstanding the quantity of its iron mines of all sorts and qualities, and the skill of its iron-masters, has not yet been able to manufacture a good blister steel with its own materials. The iron for its cementing works is imported from the North, Prussia, Sweden, and Russia. An enormous transportation trade with the North Sea has made Hull the great entrepôt of the iron necessary to the steel works of the country. From this port a quantity of vessels of all dimensions distribute the metal in great quantity at New Castle, at Birmingham, and in smaller proportion at London; but the great bulk of it goes by canal to Sheffield, the centre of steel manufacture.

426. A long use of the same products from certain iron works, and a confidence accruing from the uniformity of certain products, uniformity which assists routine and indolence so well, have given such a preponderance to some works, that they rule the market and keep their iron at very high prices. These bars have marks peculiar to each works; but,

although these marks are very numerous, the most prized are comparatively few, and are represented here in the order of their celebrity and true value:—

Fig. 16.

L (L) (Hoop L), Danemora, in Sweden.

(G L).

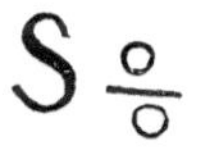

S (Double bullet).

(G F).

(Gridiron).

(J B).

(Stein-Buck).

(C and crown).

C C N D (C C N D) Russian works of Demidoff.

427. Sweden furnishes Europe with a large portion of the iron used in steel manufactures. The

country where the iron works are situated occupies the centre of the kingdom, from the terminus of Norway to the Gulf of Bothnia; its limits north are determined by a line extending from Glommen to Soderham and crossing the Lake Siljan; the limit south is the line of the 59° of latitude north. It comprises sixteen thousand square miles. The iron most esteemed for the manufacture of steel comes from Danemora, which mines are near Upsala. England imports from there 3000 tons of iron, which are employed at Sheffield and other manufacturing towns. The yearly production of Sweden is 70,000 tons of bar iron.

428. The irons of Demidoff have, for a long time, enjoyed a great reputation in English works; but their importation has greatly diminished since the death of the owner. This celebrated manufacturer died, a few years ago, in a duel with a German nobleman. It is said that his servant, having resolved to avenge his master, pursued his adversary and killed him soon after the murder of his master.

II.

Files.

429. Good files are made of steel, but it is readily understood that they have different qualities as they have different shapes. The best are those made from bars forged under the hammer, because the texture is more compact than when the metal has

been rolled. For flat files the bar is drawn by percussion to the proper dimension and shape. This is easily done. After this drawing, the head forger and his helper take hold of the piece and dress it completely before the blank is separated. The hammer of the forger has a particular shape, like a truncated cone, the base of which is flat, and covers at each blow a large portion of the file, which expands and is smoothed all at once. The hammer of the helper is large and heavy.

430. Triangular and half round files are forged in a die, a kind of swage tool fixed in the anvil; or better yet, in rollers, the grooves of which correspond with the shape of the file. The finishing is done with the hand hammer, taking care to have the tang and the other end tapering regularly, particularly with triangular files which terminate in a point. Round files are made in a similar way, and small flat files are cut from steel plates.

After this forging, and while the steel is yet soft, it is marked with a strongly hardened stamp.

431. The next figure represents a small circular forge used at St. Etienne, specially for the manufacture of files: one is enough for shops of some magnitude. Its description will give an idea of the numerous advantages it presents.

Standing in the middle of the workshop, its diameter is about 1.50 metre, and its height 0.80

metre. The inside is spherical, and at the bottom some fire-bricks are so arranged that four, six, or

Fig. 17.

eight tuyere holes will receive the blast from a central and underground pipe, B. A conical mantlepiece of stout iron plate, lined inside with bricks, covers the forge fire, at a distance of 0.10 metre from the top edge. Strong iron bars attached to the girders, keep it, as well as the metallic flue on the top, suspended.

The forge receives the blast from a ventilator, running continually during the working hours, and maintaining an intense heat.

All around the brickwork A, is an iron railing, C, for supporting the tongs. These are sliding tongs, where the piece of steel (*blank*) remains fixed during the entire heating. By resting the tongs upon the railing and the brickwork, all the blank is exposed to the fire.

The workman is able to watch the temperature without exceeding the proper degree of heat. The action of the air is less than in ordinary forge fires. There is more regularity in the manufacture, more economy of time, materials, fuel, and above all, a uniform and intense heat.

432. The files must be annealed in the forge before cutting. We have given a few precepts for this annealing, which should not be mistaken for the tempering after hardening. For annealing, the blanks are piled up in a brick furnace having a fire underneath, and so built that an intense heat can reach all parts of the pile. The fire generally lasts twenty-four hours, and, during that time, it is kept as regular as possible. When the workman supposes that the blanks have softened enough, all apertures for air are closed, the pile is covered with hot ashes, and the whole allowed to cool. This method, which is generally employed, has the defect of not preventing the oxidation which injures the steel. It is preferable to anneal the blanks in boxes or chests perfectly closed, which, it is true, will not be heated so rapidly, but will prevent all access of air.

433. After this annealing, the blank is made clean and accurate by filing (*stripping*), or grinding. The

latter method is that most generally employed; the former is in use only in a few places in Lancashire, where the long practice and the incontestable skill of the workmen give to the products of these works a true superiority. This explains why a process evidently more slow and costly, has been continued.

434. The grinding of the blanks possesses the great advantage of being rapid, and of dressing sufficiently, if the grindstone is thick enough. At St. Etienne and Sheffield the grindstones receive their motion from small water powers. In small shops, the stones are made to revolve by hand or foot. This latter method, which is free from danger,[1] is also more slow and costly.

The file is presented flat or upon the side to the grindstone, which always turns towards the workman. It is only at the last moment that the grinder

[1] Wm. Durham has computed that the velocity of an ordinary grindstone is one-fifth of that of a cannon-ball. Therefore, it often happens that fragments which burst from the surface are projected to a distance, wounding the men. Such accidents are generally attributed to centrifugal force; but this explanation is far from being satisfactory, and does not explain the explosions which sometimes take place, breaking the grindstones into many pieces, killing men inside and outside the works, and throwing down parts of walls. There is something more than centrifugal force, which would not be over 80 metres per second, at a maximum; the violent explosion, the scattering in every direction, the complete breaking and pulverization of the grindstone, all would tend to attribute the phenomenon to a powerful force, electricity, for instance, developed by friction.

is able to give the last finish and the proper uniformity.

435. After grinding, the blank is cut. The workman, sitting upon a low bench, has the anvil between his knees. The blank is put upon a block of lead, and is firmly secured there with a leather strap pressed down with the foot. The workman then presents to the blank and at a certain angle, a small chisel, very hard, well sharpened and dressed, upon which he strikes with his hammer rapidly and uniformly. The chisel, in moving, makes parallel and equidistant grooves. By practice, he preserves the three conditions of depth, parallelism, and equidistance. In this way are made the *single cut* files employed for bronze, brass, copper, and other soft metals.

For filing iron and hard metals, another cut is given to the file, by striking the chisel in a direction diagonal and oblique to the first indentations. The result is the *double cut* and the production of a quantity of sharp teeth.

The files for wood or rasps are not cut with a wide chisel, and the indentations do not occupy all the width of the file. A triangular-pointed chisel struck obliquely to the face of the file, raises a small tooth at every blow. Their number and their height correspond to the fineness of the rasp and to the work it will have to perform.

436. In various countries many attempts have been made to replace, by a regular machine, the uncertainty of manual labor. But none of these machines having been completely satisfactory, we will abstain from mentioning them, and also of a process by electricity, for which we obtained a patent in 1857.

437. After cutting, the steel of the file is too soft; and in order to perfect the instrument, it must be hardened according to the principles already explained.

438. The heating furnace for hardening, seen in the next figure, is very primitive in its construction,

Fig. 18.

and might be advantageously modified. It consists of a fireplace where pit coal is employed, and of two or three tiers of muffles. The whole height is 2 metres. The grate bars are 0.60 metre above the floor, and receive the fuel by a lateral opening, in front of which is a platform for the coal. This platform being a little above the grate bars, the workman has only to push the fuel, when the fire requires a fresh supply.

439. The first tier of muffles is 0.40 metre above the grate bars; the other tiers are 0.15 metre distant from each other. The whole apparatus (ash pit and foundations excepted) is not over 1.50 metre high, so that the workman has no difficulty in watching the interior of his muffles.

These muffles are so disposed, that the upper ones correspond to the empty spaces between the lower ones, as may be seen in the figure. The flame therefore, by raising, envelops every muffle alternately, to the last. The depth of the furnace is 0.80 metre; the front wall has round apertures left in it for receiving the muffles and removing them easily, in case of rupture. The back wall has also supports for them.

The muffles are made of the same refractory clay used for casting crucibles, and require the same care in drying and annealing. Their length is 0.80 metre, which will vary with the dimensions of the furnace. Their diameters outside, and in the clear, are 0.12

and 0.09 metre, leaving a thickness of 0.015 metre. They are closed up with conical plugs having a hole at the top, in order that by inserting the tongs in it, the workman might remove and again replace them with great rapidity.

440. The purpose of the muffles, in heating the files previous to hardening, is to protect them from contact with the air, thus preventing an oxidation and decarburization of the steel. We shall see hereafter that this is not the only precaution taken. Another advantage of this apparatus, is to allow the workman not to exceed the color and the proper heat for hardening. He finds himself thus working under regular conditions; his observations and operations are, therefore, marked by a constant regularity.

Back of the workman, and leaving him only 1.20 to 1.40 metre of free space, is a large trough about 2 metres long, 0.60 metre wide and 0.60 metre deep, for the immersion of the files. Back of this trough, and parallel to its length, is an elevated bench upon which the workmen sit who make the dipping; their feet are level with the top of the trough. The water used is at about 28° C., and the temperature is kept at this point by the heat of the files, and by the circulation of the air. Therefore, the troughs must be of a certain size; were they too small, the dipping of the files would elevate too much the temperature of the water, the hardening would be irregular, and some files would be harder than others. It is precisely a constancy in the degree of hardening, which makes the reputation of the manufacturer. It would be more rational to keep the water at a certain degree of temperature by a supply of cold water, and the use of a thermometer; but the custom is to leave to the judgment, and sometimes to the routine of the workman those operations, the importance of which should require all the attention of the master.

441. The files which are to be hardened are smeared all over with a magma of soot and salt in the following proportions:—

Soot	4 parts.
Chloride of sodium (common salt) .	1 part.

This mixture, with a sufficiency of water, is spread over the files; the object being to protect them completely against the air already in the muffles, or that which might enter, each time that the workman takes the plug out to watch the operation.

In this case, the soot acts like charcoal in the cementation of iron; it will return to steel the small quantity of carbon which may have been lost by an accidental oxidation. As for common salt, we understand its use only as the continuation of those secrets we have already mentioned. It may, however, give more firmness to the covering; and, as a part of it will be transformed into carbonate of soda[1]

[1] More likely silicate of soda; the soot containing always some sand.—*Trans.*

with the help of the soot and of the heat, it will be useful in the scouring of the files, after hardening.

442. The files, well covered with this paste, are carefully placed in the muffles, and the plug inserted. As soon as a cherry-red heat has been reached, the workman, with his tongs, turns the files upside down, exposing to the hottest parts of the muffle those which require most heat; and, as soon as the maximum of cherry-red heat is obtained, each file is taken out separately by the head workman, who plunges it, about 1 inch, into the water, and delivers it with the tongs to his helper. The head man then takes another tongs and returns to the furnace, while the helper continues to dip the file slowly and gradually into the water down to three-fourths of its length; at this moment the tongs is opened, and the file falls to the bottom of the trough. The tongs is then returned to the head workman, who hands another with a file in it.

The important part of the hardening, is the degree of temperature of the file: if the muffle is at too dark a heat, the file will be soft; if the heat is of too light a red, the steel becomes brittle, and the teeth break by the smallest strain; a broken tooth causes the next one to break, and so on, until the file is shortly useless. The hardening should be such that the teeth will crumble partially instead of breaking off entirely; there will always be a sharp angle left, which will abrade the piece to be filed, and the file will be serviceable even, when over one-third of each tooth has been worn off.

443. The workman will ascertain the proper color of the piece much better when the light of the workshop is not too bright, otherwise he would not be able to judge with certainty the degree of heat, even in the muffle, and would be liable to exceed the proper degree. At any rate, it is indispensable that the light should be equal, and, on that account the workshop should receive the light from the north. All these precautions seem of little moment by themselves, but they are parts of the success, and should not be neglected.

444. It is evident that the success in hardening files depends entirely upon regularity in heating, and an equal temperature of the water. The difference in hardening is due to the difference between the respective temperatures of water and of the heated piece. The temperature of water is easily regulated. Therefore all the attention of the workman ought to bear on the proper color of the metallic piece; a practised eye and judgment have to make up for mathematical precision. This explains how difficult it is to find good help for hardening, and why this kind of work demands such high prices.

445. The furnace for hardening which we have represented, employs a staff of two head men and

three helpers. True it is, they are not constantly occupied there, and while the heating is going on, the helpers scour the files.

446. The immersion of the files, one after the other, is done without haste, and the rapidity is in proportion to their temperature. By dipping too rapidly, the water boils by its contact with the file, and the hardness is less. In some cases this is enough to make it warp. On the other hand, all the body of the file must have sufficient time to cool uniformly; therefore, it should not be removed too rapidly. If this happens, the corners of the file will be hardened properly, while the *swell* of the file will not be so.

Generally, the steel becomes cleansed during the immersion; a good color will appear when the steel is properly carburized and homogeneous. It may also present a similar shade when the difference of temperatures of water and steel is too considerable, when the immersion is too rapid, and when it has been forged too hot.

447. We have shown in the chapter on hardening, that some pieces rather thin, after being taken out of the water, would warp and present in their length cracks, which are true breaks. This occurs frequently in hardening files. We have said also that this distortion was often due to the bad working of the blacksmith, and we gave as a proof that it rarely happens with skilful men. In regard to files, we may

add that their annealing made at too high a temperature is another cause of their distortion and breaking. Some men also dip the file obliquely into the water, beginning by the point. If done too quickly, the piece of steel meeting some resistance from the water, has a tendency to bow and to warp.

448. When a file is warped, there are several ways of straightening it again, which are employed according to the shape and the resistance of the file. Sometimes the workman profits by the remaining heat to strike it gently with a wooden mallet until perfectly cooled. Sometimes the file is put upon a block of lead or of wood, the curved part above, and pressed down by the full weight of a man. In some cases, the file is squeezed in presses of various shapes, which, in some workshops, are on top of the reheating furnace.

449. The fine double cut files of French manufacture are justly celebrated for the uniformity of their teeth and their fine cut. In this respect we are superior to our British friends; but they work their coarse-cut files so well, that they are without rivals in Europe. Let us add also that their materials are superior and better chosen than ours, and that many of our file manufacturers are of a remarkable bad faith. Not only are many files made of ductile iron, and case-hardened afterwards, but the

bars are so imperfectly worked and drawn, that after a short use, furrows will appear and cause these files to be rejected. When will these French manufacturers feel that it is as much to their dignity as to their interest to produce good articles, and to cease the manufacture of these goods *for exportation*, *i. e.*, for deceiving foreigners? Once deceived, they do not call again, and we complain of a decrease in our exportations!

450. Although there is no instrument which requires in its use less dexterity than a file, it is rare to meet a good filer, combining what workmen call *the eye* and *the hand*. The secret[1] consists in laying the file so, that the surface of the filed piece be perfectly level, without any of those irregularities left by an ordinary workman. Some machinists not only avoid making a convex surface, but also any concavity which might be expected from a stroke given in the direction of a circular segment; and this not only upon large pieces, but also upon small works which require a great nicety in their adjustment. When very large pieces are to be filed, the surface is heated red, and a large file called a *float* or *rubber* is worked by two men the same as a crosscut saw. Such files are nearly one metre long, and have handles at both ends.

451. The files employed by locksmiths are gene-

[1] Is this the "secret" or the "aim"?—*Trans.*

rally double cut. The rough work is done with bastard files, the finish with smooth ones. The latter are worked obliquely forwards and backwards. The finest files often require to be oiled, although oil prevents somewhat their *take*.[1] With rasps oil should be avoided.

452. A file should be of an uniform light gray color, and the end lighter still. To prevent the tang from breaking, it is strongly tempered in a lead-bath.

453. The trial of a file in order to ascertain if it possesses the same degree of hardness in its whole length, is made upon a piece of hard steel (not hardened), and pressed in a vice. This plate of steel is about 0.008 metre thick for large files, and 0.005 metre for small ones. The oxide of the surface is first removed with an old file; taking then a new one and laying it on the edge of the steel plate, it is drawn backwards lightly, and with a certain twisting motion. How the file "takes" is then felt, and also its hardness, by the resistance encountered. The same operation is performed at several points on the

[1] Certain essences, and among them, dead oil from coal tar, will make old files "take" as well as new ones, but will not do for smoothing. The best use of dead oil, notwithstanding its bad smell, is for taking the grease from files, pieces of machinery, &c. It is employed for this purpose on several French railways. Dead oil from coal tar is very cheap, and should be free, as much as possible, from naphthaline.—*Trans.*

file; and then, by giving lightly a full stroke as if the steel plate was to be filed flat, the avidity of the teeth is ascertained.

454. All files of a similar shape must give the same results at the trial, if their color is the same. A file of a gray color at the ends, and white at the bellied part, should be rejected, because the teeth would break rapidly at the white portion, which has been overheated.

455. Half-round files are generally harder on the convex part than on the flat. This is due to the difference of cutting; the spaces between the indentations, on the convex side, being less than that on the flat one, the cooling action of the water is more sudden.

It often occurs that half-round files are of good quality on the flat and on four-fifths of the convex part, while the thickest part soon becomes bare, and may be filed off. This defect is due to a certain quantity of oxide left by the forger, which has not been removed by filing or grinding.

456. The trial of bastard files requires a little more caution, because the indentations are finer and more delicate. The steel is generally of a better quality, has been dressed with more care, and is of a finer shade of color.

However, some bastard files will be found harder

on the edges than on the middle of the flat; in such case, the color is deeper in the centre than on the edges. This is due to an irregular heating. Sometimes, the deeper color of the centre appears like a blackish vein; it is iron fraudulently put in. When obliged to give back files weighing a certain proportion to the steel bars given to them, some workmen, in order to cover the loss, will make it good by adding to the files some of the iron employed for binding steel bars.

457. Smooth files are tried in the same way upon the smaller of the two trial bars of steel. These files are apt to get injured in the trial, and it happens very often that when filing flat, one grain of hardened steel has become stuck in the trial bar, and produces a furrow in the whole length of the tool on trial.[1]

[1] Before being ready for sale, the files pass through the ordeal of several washings in clear water, in order to dissolve all the salt which may remain upon them, a scouring with a brush and coke dust, a last washing in lime-water, drying, and, lastly, while yet hot, an oiling with olive oil mixed with a small quantity of turpentine.

Instead of salt and soot, a paste, for covering the files previous to heating, may be made of beer grounds, salt, and damaged flour. Some persons add charred leather pulverized, or a small quantity of yellow prussiate of potassa, which is a very active carburizing reagent, and, therefore, should not be used with steels already highly carburized. The teeth would be too brittle. —*Trans.*

III.

Steel Wire.

458. *Steel wire drawing* is exactly the same as iron wire drawing; the only difference is in the nature of the materials, which, in this case, require more frequent annealings and a less velocity when passing through the draw-plates.

459. The object of wire drawing is to reduce metallic bars to very small dimensions. Formerly, when the hammer alone was employed, the very fine sizes could not be obtained; however, wire fine enough for cards was made at a very high price. The rollers which took the place of the hammer, could only make round wire of 0.003 to 0.004 metre diameter, it being very difficult to have the two grooves to coincide perfectly. The invention of draw plates resolved the problem, and the remaining difficulty was to find out the proper means for hardening the holes, and for easily drawing the iron or steel rods, whose diameters exceed those of the holes in the draw-plate.

460. In general, the draw-plates are made of hardened steel, supported against another strong plate of perforated iron. Their dimensions also vary with the calibre of the wire, and they have conical holes of gradually decreasing dimensions, the last being the finishing one.

461. These conical holes would be soon worn out if they were made of a soft material. To obviate this, a very hard steel is employed called *Savage steel* (acier sauvage), and made out of pig metal. It contains an excess of carbon. That the steel by drawing will become harsh, is not of much consequence, because it can always be annealed under a layer of clay. The holes of draw-plates are punched while hot, with punches of different diameters, on account of the conical shape of the hole. My father, in the *Manuel du Maître de Forges*, thinks it would be a great deal better to drill these holes in the cold plate, which might thus be finished in one heat.

462. Hardness, roundness of the holes, and precision in the calibre, are the three conditions required by draw-plates.

The hardness, besides the economy in the long wear of the conical holes, gives more regularity to the wire, and allows a more rapid drawing. A Mr. Brockeds or Brockedon had proposed to make the holes of diamond or other gems, and this could have been done, had not the price been so excessive. Recent trials in the manufacture of artificial gems, permits us to anticipate that in a few years it will be possible to employ them. With harder draw-plates, the wires will be more regular, and their fabrication less expensive.

463. The wire coming from the rollers has its end hammered or filed to a point, in order to facili-

tate its passage through the hole. It is then caught by a pair of nippers or *dogs*, and drawn by hand or by mechanical power. In the latter case the wire is attached to the circumference of a revolving reel, receiving its motion from the machinery. Another reel, on the other side of the draw-plate, supports the wire to be drawn. This disposition obviates the marks left by the dogs, and is altogether preferable in rapidity and regularity.

The rapidity of motion of the reel is increased in proportion as the diameter of the wire decreases.

464. It has been discovered that, when the wire has been covered with a very thin film of copper, such as might be produced by dipping it through a somewhat acid solution of a copper salt, it passes through the draw-plate with great facility. However, the coating of copper does not long resist a succession of drawings and annealings.

465. By passing through the draw-plate, the steel becomes brittle and hard. In order to give it some softness, it is annealed at a dark red heat. Iron wire must be drawn twenty-one times, and annealed four times; whereas steel wire must pass through twenty-four holes, and be annealed each time, before it is reduced to the size of a knitting needle.

466. Steel wire also requires to be drawn more slowly than iron; being harder, it offers more re-

sistance, and if the operation were performed too rapidly, it would break or lose its tenacity.

467. When annealing steel wire, it is very difficult to prevent some oxidation by which small scales are formed on the surface of the wire. These scales, being very hard, would be destructive to the draw-plates; therefore, they are removed in a pickling bath, or by striking the wire with a wet wooden mallet. In some works, the coils of wire are plunged into a clay bath, dried and annealed.

468. The annealing furnace is a kind of reverberatory furnace, in the centre of which are cast or wrought iron bars, supporting the wire in the middle of the flame. This grate will receive about 300 kilogrammes of wire. The annealing is always made on the same quantity of wires, without regard to their diameter, except that the larger ones are in the hottest part, and the whole equally heated at the same time.

469. In the wire manufactory situated at Aigle, another apparatus is employed, and gives good results. The furnace is cylindrical, 1.60 metre diameter, 2.80 metres high, and with a parabolical dome. The inside contains three horizontal grates; the middle one receives the fuel, and the lower one is used as an ash-pit. The upper one carries a hollow annular cylinder made of two parts, put one within

the other and luted. The flame plays on the outer and inner sides, and the wire, put into the annular space, is protected against contact with the atmosphere. The larger cylinder is 1.40 metre in diameter, the smaller, 1 metre only.

These cylinders may be placed in or out of the furnace, at will, and are carried upon iron rails. When one of these double cylinders is taken out of the furnace, it is replaced by a new one.

Care should be taken not to open the cylinder immediately, because the steel wire is red hot, and would become oxidized. Each cylinder receives 125 kilogrammes of wire.

When all is cold, the wire is straightened by being rolled around another reel, or by blows from a wooden mallet, or upon a *riddle*. Lastly, its calibre is ascertained, in order to determine its commercial number.

470. In the month of March, 1858, Mr. S. Fox took a patent for straightening, hardening, and tempering steel wire in one operation.

He employs two horizontal furnaces, 1.60 to 2.80 metres distant from each other, at the same level, and in a straight line. This allows three distinct places: that of the first furnace for heating the wire directly from the reel; a space between the two furnaces for the hardening trough; and that of the second furnace for tempering the metal at a low temperature.

The bed of the first furnace is 1.10 to 1.40 metre long, and has, at its two extremities, two apertures for the passage of the wire.

The second furnace is arranged similarly. The wire is unrolled from a reel near the first furnace, and wound upon another reel at the end of the second furnace.

The wire passes through these apparatus, and is directed in its motion by three cast iron guides, having a length of 1.20 to 1.50 metre, a width and a height of a few centimetres.

Each of these guides is made of two pieces, in which a certain number of parallel grooves have been cut for the passage of the wire, and that with such accuracy as to have them coincide, when the two pieces are brought together.

One of these guides is put into each furnace, and projects out of the apertures for a length of 0.10 to 0.15 metre.

The third guide is put between the two others, and a supply of cold water runs through it.

471. The three guides are on a straight line, and the operation is conducted as follows:—

The wire, leaving the first reel, enters the first guide and becomes heated; it is hardened in the second one, where cold water is running; and lastly, it is tempered to the proper degree in the third guide, which is moderately heated.

472. This process has the advantage of straightening the wire in much less time than by the ordinary method. The usual way consists in passing the wire upon a board (*riddle*) where pins sloping in opposite directions are fixed. By this zigzag friction, the opposite bends neutralize each other, and the wire becomes straightened.

473. The qualities sought for in iron wire are ductility, extreme flexibility, and facility of being bent several times, in every direction, without breaking or showing flaws. The conditions for steel wire are different; it must be tenacious, that is, able to resist, without breaking, a certain tensile strain; but also it must possess a certain elasticity, which is found principally in hard steel.

474. The callipers or gauges employed for measuring the diameters of metallic wires are not uniform in every country; each manufacturer has his own peculiar gauges.

475. We have already said that the conical holes of the draw-plates were punched or drilled. These two operations leave the inner surface too rough for neat working. Therefore, it is proper to smooth the holes, by presenting both sides of the draw-plate to a brass cone or grinder fixed on the spindle of a small lathe. The abrading powder is emery, &c. The ridge produced should be without sharp angles,

and nearer the side on which the metal enters. The principle of drawing does not consist in scraping, but in pushing the metal back as in a wave.

In order to reduce the friction and to leave a smooth and finished surface on the wire, certain lubricating substances are sometimes employed, such as beer grounds, starch and water, &c.

476. The uses of steel wire are many; but their principal employment is for the manufacture of needles, and of fluted or grooved wires.

Grooved wires are used for several purposes, among them the inclosing of spectacle glasses. Fluted or pinion wires are drawn through a certain number of holes, and annealed each time; but the last drawing is made through a carefully made draw-plate, in order that the wire may have, in its whole length, a certain number of ridges and furrows representing the cogs of a wheel, when the wire is transversely cut. The hardening is done only when the piece is entirely finished and fitted.

477. Watch springs are made of steel wire, flattened between rolls or by the hammer, and finished upon a grindstone. They are hardened afterwards, and tempered blue. The labor and the numerous operations to which this little product has been submitted, enhance its value to a surprising degree: half a kilogramme of iron, which scarcely costs three

cents, produces 700,000 springs, worth over 90,000 francs ($18,000).

478. For a long time, Prussia has controlled the manufacture of steel wires for musical instruments. This may be due to the antiquity of the wire drawing industry in Germany, the first and best improvements having been made in that country. Historians assert that the first wire drawing bench was made at Nuremberg, by Rudolf, who also invented the pliers.

IV.

Needles.

479. The steel wire employed in the manufacture of needles must be of excellent quality, very hard and flexible; after being hardened it must break readily, without bending, between the fingers. Its diameter must be uniform and assorted to the sizes of needles.

480. The wire is cut with strong hand shears in lengths sufficient for two needles. A certain number of these pieces make a cylindrical package of about 0.10 metre in diameter, inclosed in two strong iron rings. After heating on a shelf, it is rubbed upon a block of cast iron with a tool called a *smooth file.* This tool is like a curved gridiron with only three bars, the iron rings or hoops of the package passing through the openings left between the three bars. By a forward and backward motion the steel

wires are straightened, and the operation is termed *rubbing*.

481. The wires are afterwards ground at both ends, upon a grindstone revolving with great rapidity. The grinder takes one or two dozen of these pieces, spreads them like a fan, and rolling them upon the stone with one of his fingers, protected by a thumb-piece of leather, points both extremities. During the grinding small particles of steel and stone dust fly all over the shop, and enter the throat and lungs of the workman, producing a dangerous sickness called *grinder's asthma*. Many contrivances have been tried for preventing this, such as covering the stone with a woollen cloth, where the metallic dust becomes fixed; a wet cloth upon the nostrils and the mouth of the grinder; a metallic mask. But the workman is always and everywhere the same; any preservative is considered by him as a restraint, which he does not care to use, and avoids. The best way is to do without his participation, and to force him to retain good health against his will. This may be effected by some magnetic sieve which will retain the metallic dust as soon as produced.[1]

[1] Metallic dust is not the only dust to be avoided; that of the grindstone is equally noxious. The apparatus which seems to resolve more satisfactorily the problem is a box inclosing the greatest part of the grindstone, and communicating by a wooden trough with a powerful aspirator. The grinder retains thus full liberty of action for his hands, and also of breathing, and of seeing.—*Trans.*

482. After pointing, the wire is cut in two, each portion being the length of a needle, put into wooden or iron boxes, and delivered to the *head flattener*.

483. This workman takes a certain quantity of the pointed wires, spreads them like a fan (the points being between the thumb and forefinger), and resting the heads upon a small anvil of polished steel, flattens them by a smart tap of a hammer. In most needle works the operation of *piercing the eyes* is entrusted to women, whose lighter hands are better adapted to that kind of work. The flattened head is put upon a small steel anvil fixed upon her bench, and struck on both sides with a small punch. The substance is only bulged out, and is entirely removed by punching the head again upon a block of lead. This operation is done by boys.

484. We must not forget that flattening the head has rendered the metal very brittle, and unfit for piercing the eye, before the needle has been annealed in a peculiar furnace.

To be certain that nothing remains in the eye, another child punches it again upon a block of lead, and after that, upon a steel anvil.

485. The next operations are *making the groove*, and *trimming the head*. The needle is fixed in sliding pincers and laid flat upon a block of wood. Then a small groove is filed both ways of the eye, in the

direction of the needle. With another file the head is rounded and smoothed.

486. In some manufactories the eyes and the groove are made at the same time by a stamping machine. With this machine 10,000 needles may be pierced in one day, while only 1500 are pierced by the hand method.

When piercing by machine, the wires of the length of two needles are not cut. The middle of the wire rests upon a raised die, while the counterpart is attached to the falling weight. Therefore, the heads are stamped, both sides at once. The place for the eyes has only been marked out; for removing the metal, the wire is put under a hand-press having two punches, which pierce the eyes at one stroke.

When the eyes have been pierced, small wires are passed through (*needles spitted on fine wires*), and the whole appearance is that of a comb. For separating the heads, the pointed ends are fixed in a wooden hand-vice, and the burr produced by stamping is filed off on both sides. A small pressure will then separate the two rows.

The last trimming is made by hand and file.

487. During the operations of flattening the head, making the groove, and piercing the eyes, the needles have almost always been curved. They are straight-

ened by rolling them upon a block of polished metal.

488. At the end of all these manipulations, the needles are gathered and inspected by the foreman, who rejects the imperfect ones, and pays the men according to the weight of the good ones.

489. The next operation is the *hardening*. Several thousands of needles are put into a cast iron tray, and covered with ashes, in order to prevent oxidation. The tray is then carried into a kind of reverberatory furnace, where it is submitted to a cherry-red heat. This temperature is most convenient for large needles, but the finer ones must be taken out before this degree is reached. When at the proper point, they are thrown into a tub of water, taking care they should not all fall at once, but one after the other, as much as possible.

490. The hardened needles are very brittle, and require tempering in order to retain some elasticity. This is done upon a plate of iron, heated nearly red hot by a fire underneath. The needles are placed in parallel lines, and constantly moved with small pincers or shovels, until a regular blue color has been imparted. They are then removed.

During the tempering, they have warped: therefore, they need straightening, which is done, one by one, upon a small anvil, and by gentle taps of the hammer. After that, they are polished and scoured.

491. For this operation, emery, rotten stone, soft soap, and oil are employed.

A layer of the abrading materials is spread over several strips of canvas; on top is a thicker layer of needles, and so on, until five or six of such alternate layers have been piled up, when some oil is spread over the whole. The pieces of canvas are then rolled, in order to form a bundle which is fastened all around and at the ends by strong twine. Each bundle holds from 40,000 to 50,000 needles.

When a certain number of these bundles have been prepared, they are put between horizontal rollers, made of wooden troughs loaded with weights, and to which a forward and backward motion is imparted by some machinery. It is necessary to regulate their motion, in order to prevent heating, which would destroy the hardness of the needles. Generally, these rollers move ten to twelve times in a minute.

The polishing lasts from one and a half to two days, and the needles come out covered with a black, thick, and greasy mixture, which must be scoured off.

492. This last manipulation is effected in cylindrical drums revolving on their axles, and partially filled with sawdust. The needles are thrown pell-

mell into it, and after a certain time, are removed, and winnowed in a copper basin. New polishing bundles are formed, rolled for eighteen hours, and again scoured. When the needles appear to be sufficiently polished, the last bundle is smeared only with oil, and remains five to six hours in the polishing mill.

493. The last scouring, in order to remove the oil, is made by forming new bundles with alternate layers of needles and bran, without emery. After remaining for a few hours under the rolling tables or troughs, they pass to the drum, and are winnowed. At last, they are wiped dry, one by one, with a rag.

494. During the polishing, ten per cent. of the needles, on an average, have lost their points, or have been entirely broken; two to five per cent. have been bent, and must be straightened. The separation and the sorting of the needles according to their size, is the last operation of all.

495. This manipulation, which is done by children, with remarkable dexterity, consists in shifting to one side all the heads, in separating the needles of different sizes, and in removing the imperfect ones.

The needles are put flat upon a table, and a dozen of them are rolled with the forefinger of the left hand, while the forefinger of the right hand, covered with a piece of rag, presses against the extremity of the needles. The pointed ends stick to the rag, and are carried to the left side. The heads are then shifted to the right side. Every motion of the finger piece carries with it half a dozen needles.

496. The eyes of the first quality of needles are sometimes drilled, in order to smooth the rough edges. The insertion of the thread is made more easy, and its cutting prevented.

497. The last operation, performed only on needles of the first quality, is a polishing upon a blue or hone stone, or buff leather. The points are sharper, and the polishing is perfect.

The needles, in quantities of twenty-five, fifty, or one hundred, are then wrapped in a paper especially made for the purpose, and which does not allow any dampness to penetrate. The packages are stamped with the mark and name of the manufacturer.

498. Very inferior needles are made of iron, instead of steel wire, and cemented afterwards in crucibles or boxes, by a process similar to those we have described for cementing and case hardening. The cementation is watched by means of trial wires, and the operation is ended when the wire, plunged into water, is clean and breaks easily.

499. Needles with gilt heads are made by dip-

ping the eye into a solution of chloride of gold in ether. The gold is precipitated immediately, and becomes fixed upon the steel. Sometimes the whole needle is plunged into the liquid, and will no longer be subject to oxidation.

V.

Steel Plate.

500. Steel may be rolled into sheets of variable thickness by a series of operations similar to those employed in making sheet iron. At present, cast-steel and steel of cementation well tilted are the only materials used; but there is no doubt that it will be possible to roll into plates natural steel, and steel from pig iron. The principal condition is, that the metal should be perfectly homogeneous, compact, resisting, elastic, and without flaws or cracks. In all these respects cast steel is pre-eminent.

Rolling steel plate requires a greater mechanical power than a similar operation with iron, because steel, being unable to bear as much heat, offers more resistance than iron.

501. The alternate use of hammers and rolls is necessary in making steel plates. The former tools are more particularly employed for preparing the steel slabs; the latter are preferable for extending and polishing.

502. When steel of cementation is employed, it is drawn into flat bars in the rolling mill, or under the hammer. These bars are then cut into pieces, the same as in working iron, piled up, reheated, and drawn again under a heavy hammer or between rollers.

503. The bundle or fagot of cemented steel is not drawn in one operation. A first slab is made and the remainder of the bundle is heated and drawn again, thus making the second slab. Both slabs are then cut out, piled up, reheated, doubled, &c.

504. It is important that the hammer should fall plumb upon the anvil, in order that the shock should be felt all over the surface, and not on a particular point. A hammer for steel plate weighs 200 to 225 kilogrammes, and its fall is 0.95 to 1 metre.

505. Not only does steel require for its working powerful forge tools, .but it is also important that the shocks should be graduated according to the wants. Steam hammers are, therefore, exceedingly well adapted to the working of steel.

506. In the manufacture of steel plates no reverberatory furnaces are employed. As far as practicable the metal is kept in an atmosphere of carbonic oxide. This is the proper way to retain the quality and the elasticity of steel.

507. The furnaces employed for reheating slabs, doublings, bundles, and plates, have no stack or chimney proper. A grate takes the place of the bed, and an arch covers the whole. The smoke issues through the working door and escapes by a metallic flue, funnel-shaped. The metallic pieces rest upon the fuel of the grate, and are sometimes covered over with burning coke or charcoal.

508. The slabs which have been doubled and reheated are drawn in the rolling mill.

509. The rolls are made of chilled cast iron, that is, of metal run into a mould of wet sand, which has the effect of producing a sudden cooling, or rather a hardening of the metal, extending as far as 0.02 to 0.03 metre from its surface, which is therefore white and very hard. The frame carrying the rolls is itself massive, and is provided with counterweights which prevent the rolls from touching each other, in order to avoid injury when the laminated plate is out. In front of the rolling mill is an iron platform, somewhat above the space between the rolls, and which supports the plates to be drawn.

510. In some steel works the slabs are roughened in a particular mill, somewhat similar to the roughing down rolls used in iron forges. The rough edges of the extended metal are again cut by strong shears, piled up, reheated, drawn, doubled, and so on, until the plate has reached the proper dimensions.

511. A certain amount of care and prudence is necessary for rolling. The distance between the rolls must be gradually diminished by turning the screws after each passage of the plate.

512. In order to prevent a decarburization of the steel, the slabs or plates receive a coating of clay, before they are reheated.

513. But, previously to this coating of clay, the metal has been deprived of its oxide by blows of a hammer; without this precaution, the crust of oxide would be impressed upon the metal. The operations of doubling and heating should not be too frequent. The doubled plates have the bend presented first to the rolls. The aim should be to give the proper width at the beginning, in order, afterwards, to draw only in the direction of the length.

514. The last rolling but one is to be made at a cherry-red heat. Immediately after the plate has left the rolls, it is dipped vertically into cold water. Another annealing and a last cold drawing finish the operation.

515. The rolls for steel plates are smaller than those for iron. However, large steel plates are employed at present in England; therefore, the rolls should be of sufficient length and of a corresponding weight. Generally, a train of two pairs of rolls is

employed: one for laminating the slabs simple or doubled, and making 25 to 30 revolutions in one minute; the other for the plates, and particularly the thin ones, revolves 40 times.

A complete system of mills for steel plate requires machinery equal to 70 or 80 horse-power.

516. For over two and a half centuries, steel plate has been employed for engraving, whether with the steel graver, or with aqua fortis (*etching*). The trials by Albert Dürer go back as far as 1510: this is the date of one of the four plates of this celebrated artist, etched with aqua fortis, which are kept at the British museum. Since that time, numerous and often useless attempts have been made for improving this cheap process of reproduction. It is only after the lapse of two centuries, at the beginning of the nineteenth, that the process has become certain and practical.

The principal advantage of steel over copper plates, is that of allowing of the printing of more copies in the press. Steel plates last comparatively much longer; but we must not conclude with doctor Lardner,[1] that a plate of decarburized and soft steel, that is, converted into iron, will be able to furnish several millions of copies. It is an exaggeration of

[1] *A Treatise on the Progressive Improvement and Present State of the Manufactures in Metal*, translated by Mr. A. D. Vergnaud under the title of *Manuel Complet du Travail des Métaux*. Collection Roret.

this savant which we are astonished to see reproduced by his translator.

517. According to the same doctor, the process of engraving upon steel would be as follows: 1. Decarburization of the steel; 2. Engraving upon the softened plate; 3. Carburization of the steel by cementation in a closed vessel.

Is it not evident, in this case, that it would be much simpler to engrave on a plate of pure iron, which could be cemented afterwards?

This is not the proper place for discussing the cabinet metallurgy of writers who make books with other books, and keep up false ideas without taking the trouble to verify them. Let us say at once, that the engravers use plates of polished steel, as they come from the mills, and without any previous or subsequent operation.

VI.

Saws.

518. Saws are made of cemented tilted steel, or better, of cast-steel.

519. The steel is first drawn into slabs by processes similar to those we have described for steel plates, and when the slabs have been laminated to the proper thickness, they are clipped with lever-hand shears. The edges are ground true upon a

grindstone, and the piece is ready for the next operation.

520. The teeth are cut out by a steel punch, acting vertically under the impulsion of a fly press, similar to those employed for stamping. At each turn, one or several teeth are cut out, and the saw, which stands horizontally upon a steel rest, is shifted accordingly. The distance of the teeth is regulated by a gauge falling into the tooth.

521. It will readily be understood that the dimensions and the shape of the teeth vary with the work required of the saws. In some saw-mills, the form is that of a rectangular triangle; pit-saws have their teeth cut so as to attack the wooden fibres at a sharp angle; hand-saws, and, generally, all those used in carpenter work, have teeth of an intermediate shape.

After punching, the teeth are finished with a file, and the irregularities are ground off.

522. After these operations, the steel is to be heated and hardened. This manipulation requires judgment and practice, because the saws do not require strong hardening. The fullest amount of elasticity is to be retained; therefore, the temperature of the heated piece should not be too high comparatively with that of the dipping liquid.

523. The saw is put flat into a furnace, and there

heated to a certain point, taught by practice. Immediately after the proper degree has been attained, it is plunged into a bath of linseed oil, cold or slightly heated. Generally, two dipping troughs are at hand, in order to cool any which have had their temperature raised too much.

Every manufacturer possesses a secret for the composition of the hardening liquid, which he carefully keeps; but the basis is oil, which, when cold, imparts the proper hardening to the instruments requiring a great elasticity; some persons add tallow, rosin, and other fatty matters.

524. The saw having been taken from the bath, most of the fatty matters remaining on it are wiped off; what remains will burn in the fire during the tempering. For this purpose, the saw is put upon a coke fire and constantly moved until the grease takes fire; it is the *blazing off*. It is then withdrawn, and straightened while it is hot.

525. The straightener takes hold of the saw with his left hand, and resting it upon a polished anvil, strikes it in every direction with a small hammer. The shrill noise produced is exceedingly disagreeable. By this manipulation the saw becomes homogeneous and perfectly elastic. The straightener must be a skilful and experienced workman, for he must determine where his hammer has to fall by the noise of the vibrating plate.

526. The next operation is the *planishing*, upon stones of a large diameter (1.50 to 2 metres diameter, and 0.25 to 0.30 metre thickness). The piece of steel is too thin and too wide to be planished with the hand like a knife; it is, therefore, stretched in an iron or wooden frame connected with the grindstone, in such a way as to allow the saw to be ground in every direction, until it is regularly planished and polished. The largest saws are suspended to the ceiling by swing rests.

527. The planishing destroys most of the stiffness of the saw. It is, therefore, necessary to restore this quality by another hammering made as before. The saw becomes elastic by a new annealing, after which it is finally polished upon a hard stone and buff-leather.

APPENDIX.

EXTRACTS FROM THE REPORT ON THE PARIS UNIVERSAL EXPOSITION, 1867.

By ABRAM S. HEWITT, United States Commissioner.

BESSEMER STEEL.

Paris, June 22, 1867.

To the Commissioners of the United States for the Universal Exposition of 1867 :—

The undersigned has the honor to submit a special report upon "Bessemer Steel," prepared under his direction by Frederick J. Slade, scientific assistant to Committee No. 6, and duly approved by the committee and ordered to be laid before the commission.

ABRAM S. HEWITT,
U. S. Commissioner and Chairman of Committee No. 6.

The Bessemer Process.

The Paris Exposition affords valuable information in reference to the capabilities of the Bessemer process for the production of all grades of metal, from a near approach to wrought iron to the hardest and finest kinds of steel. A comparison of the specimens sent from the various countries shows that the quality of the metal produced depends chiefly upon the nature of the raw materials used, and accordingly it is only in those countries where the very best ores and purest coals are employed that we find the finer grades of steel produced.

It will, perhaps, be most instructive, therefore, to examine the manner in which this process is conducted in each country separately, and to trace, if possible, the relation between the nature of the finished products and the materials and modes of working employed in their manufacture. We begin naturally with

ENGLAND.

The iron almost exclusively employed in England for the pneumatic process is obtained from the Cumberland district, and is derived from red hematite ores. Dr. Percy, in his well-known work on metallurgy, gives as the analysis of two specimens of these ores:—

	I.	II.
Sesquioxide of iron	95.16	90.36
Protoxide of manganese	0.24	0.10
Alumina		0.37
Lime	0.07	0.71
Magnesia		0.06
Phosphoric acid	trace	trace
Sulphuric acid	trace	trace
Bisulphide of iron	trace	0.06
Ignited insoluble residue	5.68	8.54
	101.15	100.26
Silica	5.66	7.05
Alumina	0.06	1.06
Sesquioxide of iron		0.19
Lime		trace
	5.72	8.30
Iron, total amount	66 60	63.25

The blast furnaces in which these ores are smelted average about fifty feet in height and fifteen feet diameter of boshes, and are in most cases open-topped, the opinion among the iron-masters being that the quality of the iron is injured by any attempt to draw off the gas. At some furnaces, however, this notion is abrogated, and the waste gases are utilized for heating the blast. Among these are the furnaces of the Barrow Hematite Iron and Steel Company, the West Cumberland, and the Wigan Iron and Coal Company's furnaces. The quality of pig produced at these latter works does not perhaps stand invariably as high as that of the Whitehaven Hematite Iron Company (Cleator), the Workington Iron Company, or the Harrington, but if there is a difference it is easily accounted for by the quality of the materials used, without the necessity of resort to the supposition of an injurious effect from utilizing the escaping gas.

The fuel used at the furnaces in the Cumberland district is the best Newcastle coke, which is remarkable for its hardness and freedom from sulphur. Dr. Percy gives the percentage of sulphur as 0.8, and of ash 4.45. No charcoal pig is made in England for the Bessemer process. The fluxes employed are a limestone quite free from phosphorus, and a portion of black shale from the coal beds, consisting of clay and carbonaceous matter, without any appreciable amount of sulphur. The percentage of iron indicated by the above analysis, viz., from 60 to 70, appears to be a fair average, and the ores are not calcined. As it is necessary that the iron should be as gray as possible, not less than thirty hundred-weight of coke are used per ton of iron produced, and a charge is about fifty hours in coming down through a furnace of the dimensions given above. The yield from such a furnace is 250 tons per week.

The blast is under a pressure of three and three-fourths pounds, and is heated to from 650° to 750° Fahrenheit. From four to six tuyeres are usually employed. No. 1 iron for the Bessemer process from these furnaces brings ninety shillings per ton at the works, and No. 2 ten shillings per ton less.

The Wigan Iron and Coal Company, Lancashire, produce an iron which is used to a considerable extent for the process, but does not rank as high as the Cumberland irons. The coal as mined would be quite unfit for use in the production of such a grade of iron, as it is materially contaminated with sulphur, but this is almost entirely removed by washing the fine coal, the pyrites settling by their superior weight, while the pure coal is carried on to receiving beds by the current of water, and the purified residuum is then converted into coke, yielding a tolerably strong product. This company have just erected a number of new furnaces much above the usual size for this kind of iron, viz., eighty feet high and twenty-four feet diameter of boshes, and these are provided with a cone and bell arrangement for taking off the gas.

Forest of Dean iron, made from brown hematite ores, is frequently used in small quantities in admixture with other irons for the purpose of maintaining the heat of the charge, which it tends to do. It is apt, however, to contain too large a percentage of sulphur to work well alone.

Another brand which is said to work well is Weardale, an iron made from spathic ores. It is unusually rich in manganese, and owes its excellence chiefly to that fact.

The following analyses exhibit the characteristics of some of the more usual brands of iron employed :—

	Cleator.	Workington.	Weardale.	Forest of Dean.
Carbon (graphitic)	4.007	3.14	3.24	3.25
Silicon	1.752	3.12	1.80	1.36
Sulphur	......	0.05	0.04	0.037
Phosphorus . .	0.049	0.03	0.19	0.000
Manganese . . .	......	0.02	1.45	

The analysis of Weardale is taken from *Percy's Metallurgy*; the others were furnished to the writer from different sources in England.

The presence of silicon in the iron causes the charge to work hot in the converter, and it is usual, therefore, to mix an iron rich in this element with others containing a less quantity, and which have a tendency to work cold and become pasty. As a rule Workington iron contains more silicon than any other in use for the process, and being moreover an excellent iron is largely used. It is, however, from the very fact of its working so hot, seldom employed alone, as it cuts the moulds badly in pouring.

Sulphur and phosphorus are the most injurious elements found in the pig, because the pneumatic process is powerless to remove them, and the quality of the steel is materially affected by their presence. An effectual means of eliminating these substances, in the process of conversion, would be one of the most valuable discoveries of the times.

It is usual among all the steel makers to mix several different brands of iron where a uniform and good quality of steel is desired, but there seems to be no definite mixture which is agreed upon as best. The principle appears to be to form the larger portion of the charge of the better brands of Cumberland hematite, and to add as correctives smaller percentages of other irons. The following will serve as examples, the first having been given to the writer by Mr. F. Preston, late managing director of the Lancashire Steel Company, and the other being from the books of another large firm :—

I.		II.	
Workington	45	Cleator	40
Harrington	40	Workington	20
West Cumberland	10	Harrington (No. 1)	15
Wigan	20	Harrington (No. 2)	5
Weardale	7	Forest of Dean	10
Forest of Dean	3	Wigan	3
	120		93
Spiegel	7½	Spiegel	6¼ or 6½
	127½		

For forgings such as axles, tires, locomotive crank shafts, etc., none but No. 1 iron is commonly used, but for rails a greater or less amount of No. 2 is added, in order to reduce the cost as far as possible.

The amount of this quality that may be used will of course depend on the character of the iron.

The iron as a rule is melted in reverberatory furnaces,

but at five works, cupolas have been substituted with apparently good results. These are:—

The Manchester Railway Steel and Plant Co.;
Messrs. Chas. Cammell & Co., Penistone;
The Bolton Iron and Steel Co.;
The Barrow Hematite Iron and Steel Co.;
The Mersey Iron and Steel Co., Liverpool.

At the latter a cupola is also employed for melting the spiegeleisen. At the first-mentioned works Woodward's patent steam-jet cupola is employed, it is stated, with a consumption of coke as low as one and one-fourth pound per hundred weight of iron. At the others, Ireland's upper tuyere cupolas are employed. These cupolas melt very rapidly, and are sufficiently capacious to hold an entire charge in the portion below the upper row of tuyeres. The size erected for a five-ton plant is seven feet in diameter, and will melt five tons of iron in three-quarters of an hour. In working, the charge is weighed when it is put into the cupola, and, as it melts, remains in the bottom till the whole has been fused, when it is tapped off into the converter. They generally require cleaning once in twenty-four hours. Of course where cupolas are used, much greater care has to be exercised in the selection of the coke, as fuel which might be used in the air furnaces would destroy the quality of the iron if burned in contact with it. The opinion among those who employ the cupolas is, that it is quite possible to find a coke sufficiently free from sulphur to yield a satisfactory result. At the Barrow works, preparations had been made to convey the molten metal directly from the blast furnaces to the converters, but after a number of trials it was found that the uniformity of the metal could not be relied on, and, in consequence, the attempt was abandoned, and cupolas erected instead, to remelt the pigs. The converters at the majority of the works have a capacity adequate for a yield of five tons of steel, or allowing one-sixth for waste, which may be taken as a fair average, for six tons of molten iron. At Barrow, however, three seven and a half ton vessels have been erected, besides their five-ton plant, and at Messrs. John Brown & Co.'s a pair of ten ton vessels have been in use more than three years. The material commonly employed for lining the vessels is ganister, a highly silicious substance, found at Sheffield. Other materials have been tried at some works, as for example, at Dowlais, with apparently great success. A pair of vessels, at the works just mentioned, had recently stood 300 blows each, without relining, and were still apparently in good condition. This is much above the average endurance of the refractory linings. The destruction of tuyeres is an important item in the expense of the process. The average life of these is seldom over five blows, and the failure of one during a blow is often the cause of considerable loss, either by damage to the vessel or by injury to the contained charge.

In the general arrangement of the Bessemer plant, very few changes have been made from that planned by Mr. Bessemer and contained in the drawings supplied to his licensees. A pair of converting vessels usually placed opposite to each other, but in some cases side by side, stand at the side of a casting pit, sunk a few feet below the general level of the floor. These vessels are mounted on trunnions, and are revolved on them by means of a rack and pinion operated by hydraulic pressure. The melting furnaces are placed in a room having a considerably higher floor level than the converting room, so that the melted metal may be run by its own gravity into the mouth of the converter, when the latter is turned down suitably to receive it. In the centre of the pit is a vertical hydraulic piston or crane, carrying at its upper end a platform, at one end of which is a ladle sufficiently large to hold the contents of the converter at the end of the operation. The platform is furnished with gearing, so that it may be easily revolved to bring the ladle over each ingot mould successively, the latter being arranged accordingly in the arc of a circle near the side of the pit, which here has the same form. The ladle is provided with a nozzle and stopper in its bottom, by means of which the flow of the steel is regulated. Two hydraulic cranes, consisting simply of vertical pistons, carrying a long horizontal jib with a rolling carriage, to which a chain and hook is attached for lifting the ingots, are placed near the edge of the pit, about opposite the centre of the converters, and serve also to lift off the various parts of the latter when required for repairs. The blast valve and hydraulic apparatus pertaining to the converters are worked from a valve stand, placed at a suitable distance from the pit, the cranes being operated by a valve directly attached to them, so that the attendant boy may the better see what he is required to do, and the whole of the manipulation of the vessels, ladles, and ingots, gives an ease of working and a perfection of control, with economy of labor, which should lead to the more general application of hydraulic power to other departments of industry in which large masses have to be dealt with. The water pressure used for the purpose is about 300 pounds per square inch. The sizes of ingots most commonly cast are, for rails, about 10 inches square, for locomotive crank shafts, ingots of a rectangular section, say 22 inches × 16 inches, and for other forgings according to the size and nature of the work, the moulds having a weight about equal to that of the ingots. At some works, the plan is adopted of testing a sample of each blow for carbon, and classifying the metal according to the result of this test. By this means much greater uniformity in the finished work is obtained, and in the present state of our knowledge of the process, this is a very necessary means to secure this end, and should be more generally adopted. The process employed was introduced from Sweden, and is exceedingly simple in its nature. It consists in dissolving a known weight of metal in the form of drill chips, or some other finely divided state, in nitric acid, of the gravity 1.2. The solution will have a brown color, more or less deep according to the percentage of carbon contained in the metal. A standard color, corresponding to a known percentage of carbon, as determined by direct analysis, is first established, and the color of the solution to be tested is made to agree exactly with this by the addition of a certain quantity of acid or water. That this, which is the readiest method of producing agreement, may be employed, the color of the standard solution must be light. The water is added to the solution in a graduated test tube, so that the exact proportion of water relatively to the original solution may be read off with ease, and if, for example, an equal bulk of water requires to be added to make the color the same as the standard, the percentage of carbon in the specimen under test must be just double that of the standard. As a solution of steel in acid would in the course of time change its color, an exact imitation of it is made by dissolving burnt sugar, and this is kept hermetically sealed for comparison. To secure a light standard color, it is not necessary that the piece of steel dissolved should contain a small percentage of carbon, but a larger quantity of acid may be used in a known proportion, say twice or three times the required amount, and the corresponding percentage of carbon will be equally well ascertained. This test is easily and quickly applied, and the variation of color being considerable, gives results sufficiently accurate for the purpose of a proper classification of the ingots according to the purposes for which they are suited.

The principal uses to which the Bessemer metal is put in England, are the manufacture of rails, tires, axles, machinery forgings, and boiler plate. The total amount produced may be judged from the fact that the quantity made per week at the works of Messrs. John Brown & Co., limited, and Messrs. Chas. Cammell & Co., limited, is stated to be 600 or 700 tons each. The number of establishments at which the process is in operation is about fifteen, and the number of converters employed upwards of fifty. The chief market is for rails, and a large proportion of the orders are for *American roads.* In England, not much ordinary line has been laid with steel rails, but on most roads those portions which are exposed to excessive wear, such as stations and inclines, are being relaid with steel. The public are already familiar with the vastly superior endurance of steel in such situations, and nothing need therefore be said here on that point.

MANUFACTURE OF STEEL RAILS.

It is usual, as already stated, to cast a 10-inch square ingot for rails. At most works, this is reheated in a reverberatory furnace and hammered down to 7 inches square. At some prominent establishments, however, this process is dispensed with, and a 10-inch ingot is taken directly to the rolls and rolled down to 7 inches. At Crewe, Mr. Ramsbottom employs a heavy cogging machine for the same purpose. This is simply a form of reversing rolls made exceedingly large, and only performing a part of a revolution at each pass of the ingot. It is stated that the rails made from unhammered ingots stand equally good tests with those which have first undergone hammering.

The substitution of rolling, of course, cheapens the manufacture, and reduces the amount of plant necessary, as well as the number of hands required. It is usual after the ingot has been brought from 10 inches down to 7 inches to put it back into the heating furnace for a short time, to bring it up to a heat sufficient to carry it through the remainder of the process. With hammered ingots it is usual to allow them to become cold after hammering, and to reheat them entirely anew, since it is not easy to regulate the heats so as to have the hammer supply hot ingots to the furnaces for the rolling mill. This, of course, involves a further additional expense in the use of the hammer. In heating the ingots care has to be taken that the heat is not forced so as to burn the steel, and ample time must be given for it to "soak." Practically about four heats are obtained in twelve hours, where with iron seven or eight could be got. When the ingots are rolled from the cast size, it is usual to provide larger furnaces and a greater number for the first heat than for the second, as the fewer and smaller ones will work off the same number of ingots, on account of the shorter time necessary to bring them to the required heat. At the Dowlais works, for example, there are seven furnaces holding seven ingots each for the first heat, and but four holding four a piece for the supplementary heating.

The usual size of rolls for steel rails of the English (80 lbs. per yard), or other pattern is from 22 inches to 24 inches diameter. In some cases, however, smaller sizes are in use, as at Crewe, and at the Mersey iron and steel works, at the latter of which only an 18-inch train is employed. These, however, are trains which were originally intended for rolling iron rails, and have been compelled to do service for steel.

The speed with rolls of the first mentioned sizes varies from sixty to forty revolutions per minute; the former extreme, however, seems preferable. The drafts on the rolls are made somewhat lighter and more numerous than for iron—say two more grooves for finishing.

At several works reversing rolling mills have been erected, to avoid the necessity of lifting the ingots in returning, and also to save time by operating on the ingot when moving in either direction. The usual plan has been to effect the reversing by engaging by means of a clutch gears running in opposite directions. This necessarily brings a severe shock on all the machinery, especially at high speeds, and in some cases where the arrangement has been introduced it is not used, the mill always running in one direction, and the rolling being carried on in the usual way. Mr. Ramsbottom has constructed and patented a reversing mill, which he uses for rolling locomotive frame plates, at Crewe, which is free from this objection. He drives his rolls by a pair of engines, resembling a set of locomotive engines in most of their details, and without any fly-wheel. These work at a high speed, and are geared to the rolls in such a manner as to reduce the speed to the required amount. The link motion is thrown up or down in reversing by a hydraulic piston, easily set in motion by the attendant, and by these means the engines can be reversed seventy times per minute, and entirely without shock. This principle for reversing would appear much preferable to the use of a clutch. The employment of a fly-wheel is not found necessary, as the engines, in virtue of their high speed, contain power sufficient to overcome any obstacles within the limits of safety to the rolls, beyond which it is better that they should stop. Mr. Ramsbottom has adopted in this set of rolls a thorough application of hydraulic power for all the operations of manipulation, and has thereby obtained great facility of working and economy of labor. Instead of the reversing principle, a steam or hydraulic lifting gear is used at some works for raising the ingot to the level of the top of the upper roll, and by many this is preferred to reversing.

The Siemens furnace is coming extensively into use in steel works for heating ingots. At present they are in operation at Crewe, Bolton, Barrow, the Mersey works, and some other places. They require a certain amount of care in their management, but yield very satisfactory results in their working. They are expensive in first cost, but in districts where coal slack is abundant they are exceedingly economical in respect of fuel, since they allow of the use of this cheap material instead of better and more expensive coal. But even where good coal must be employed in the gas producers, the utilization of all the heat produced by combustion renders the saving of fuel very considerable as compared with the ordinary reverberatory furnace. For steel an excessively high temperature, such as is required for some operations, and which alone the Siemens regenerators are able to give, is not necessary, and where much steam power is required it may be quite as economical to employ the waste heat from the furnaces for heating the boilers as to pass it through regenerators for the purpose of heating the incoming gases for the furnaces themselves. In such a case as much and more expensive fuel might be required for generating steam under independent boilers as would be saved at the furnaces by the use of the regenerators. In this connection may be noticed a plan that has been adopted at the Bolton works with good results, viz., the heating of boilers by gas drawn directly from the gas producers. This, of course, gives the same economy in respect of the use of slack as already referred to. Where sufficient steam is already obtained or is not required at all, the regenerative furnaces are of undoubted advantage. Mr. Webb, at Bolton, states that it is still an open question with him whether it is preferable to heat his boilers, as already mentioned, by gas, or to place them over furnaces fired in the ordinary way with coal.

The sawing, straightening, and punching of rails are conducted in general as in America, with the exception that a single saw, or a pair side by side, instead of two separated by the length of the rail, is used. The length of the rail is regulated by stops on the carriage, one end being sawed off and the rail then passed along on the friction-rollers in the carriage till it reaches the stop, when the other end is cut off. The use of a single saw, it is claimed, enables the cut to be made at the most suitable point, as indicated by the appearance of the end, and also gives greater facility in varying the length of the rail as required for different orders. At Barrow, the rollers in the saw carriage are driven by friction gearing from the saw engine, so that the rail is passed along automatically; the carriage is also drawn up to the saw by a number of racks and pinions at intervals along its length driven in a similar manner.

At some works, severe tests are adopted for ascertaining the quality of rails, and until more accurate knowledge of the nature of the Bessemer ingots is obtained some such tests would appear to be very necessary. The usual method of procedure is to place a rail from each lot made from one mixing of metal, on supports three feet apart, and let fall upon it midway between them a weight of one ton from heights varying from ten to thirty feet, and observing the deflection produced. It is considered that good rails should not break under this test, though they may bend considerably where great height of fall is employed.

The use of steel-headed rails is a point of great importance, but one on which at present little that is conclusive can be said. They have been made to a considerable extent at the Crewe works of the London and Northwestern Railway Company for use on that line, and Mr. Webb (formerly of Crewe) has patents for forms and materials of piles for their production. One of the points which Mr. Webb claims is interposing a layer of puddle bar between the steel face and the fibrous iron, for the purpose of making a more gradual transition between the crystalline and fibrous metals, and thereby securing a more perfect union in the successive layers. The same thing has been done for many years in the United States. In the Exposition, specimens of steel-headed rails of French manufacture are shown, which have been struck on the top of the head with a steam hammer, cracking vertically through both steel and iron, and buckling up the web without any appearance of separation between the steel face and the iron beneath it. Although the specimen gives no evidence of being a selected one (the line of the weld being plainly marked on the external surface), yet it is clear that no such test can decide a question which can really only be properly solved by experience under the conditions of regular working. A sudden blow may be incompetent to produce effects which may follow prolonged and irregular hammering under the wheels of railway trains. While, therefore, steel-headed rails cannot be pronounced an absolute success, there is every reason for prosecuting the experiment, and reasonable grounds for anticipating a perfectly successful result.[1]

As the production of rails is at present the largest branch of the Bessemer steel manufacture, the disposition to be made of the crop ends becomes a question of immediate importance, and that to be made of the worn-out rails one of future moment. As the metal, when it contains any material proportion of carbon, is unreliable

[1] Experiments made in the United States, after a trial of two years, have demonstrated that a perfectly sound weld of the steel to iron can be secured in the head of the rail.

when welded, it is not so easy to decide to what use the large amount of ends sawed off from the rails shall be put. At present it must be admitted they are rather a drug in the market. When an iron that works hot in the converter is used, a certain quantity of these ends may be remelted in the vessel without injury to the steel. About four hundred weight per charge of five tons is considered admissible at the Dowlais works, the scrap being first heated to a red heat in a furnace placed near the vessel, and thrown into the latter before running in the molten iron. It is difficult, however, to dispose of the whole amount in this way. As large a portion as possible is sold to the Sheffield crucible steel makers, who remelt them, and sell them at a greatly advanced price. At some works, again, they are rolled into small plates, and in this form they may be used for the manufacture of plough shares and other kindred objects; or in some cases they may be rolled and drawn into telegraph wire; it would be impossible, however, to make fine sizes of wire from them. If the difficulty of disposing of the steel scrap is to continue, it forms another argument in favor of steel-headed rails, since these, when worn out, would contain but little steel and could be readily piled and re-rolled, the pile being so arranged as to bring the steel in the least vital parts of the rail in case its presence should lead to any unsoundness of the welding. It would appear, however, that an adequate market for old rails could be formed by rerolling them into the form of bars for machinery and other purposes, for which, by reason of their superior strength, they should be more valuable than wrought iron.

MANUFACTURE OF TIRES.

Next in importance to the manufacture of steel rails is that of tires for locomotive and railway carriage wheels. Four years ago it was attempted to weld these up, as in the case of iron from straight bars, but the unreliability of all tires so made was soon apparent, and the attention of manufacturers was directed to discovering some practicable means of producing them without welds. With the exception of the form of the ingot cast for the purpose, the mode of manufacture adopted at all the English works has attained a remarkable degree of uniformity. Mr. Ramsbottom casts his tire ingots in the form of a truncated cone, a usual size being two feet diameter at the bottom, six inches diameter at the top, and thirty inches height. This he hammers on its ends and sides till it assumes the shape of an ordinary flat cheese, with a thickness of about twelve inches. Another heat is then taken on it, and it is then placed under a steam hammer furnished with a pointed conical tool, and by successive blows with this on both sides a hole is forced through the centre of the disk, and this again expanded as the hammering proceeds, till the upper part of the tool, which is flat, comes down upon the tire and consolidates the metal by reducing its thickness. A third heat is then taken, and the ring so formed is placed over a stout beck projecting from the inclined side of an anvil, which maintains the ring in such a position as to give a suitable bevel to the outer face when struck by the hammer, while at the same time its diameter is considerably increased by the operation. After this third hammering it is ready for the rolls, and a fourth and last heat is taken for that purpose. Mr. Ramsbottom holds a patent for the method of punching the tire blocks by a sharp-pointed conical tool without the removal of any of the metal. The form of rolling-mill employed by Mr. Ramsbottom is exceedingly complicated, and is the only one of its kind, as far as the writer is aware, which is in use in England, unless it be at the works of the patentee, Mr. Jackson, at Manchester.

At Mr. Allen's works, Sheffield (H. Bessemer & Co.), the cheese-shaped blocks are produced from an ingot of the ordinary square form, this being cast sufficiently large to form a number of tires, say four, and then hammered round and cut up into sections, each of a weight suitable for one tire. The central hole is punched by flat-ended punches about eight inches in diameter at the lower end, and perhaps nine inches above, driven in from both sides successively, and knocking out a circular disk about two inches thick as scrap. The blocks used with this process are of less thickness, say seven inches. The hole so formed is slightly enlarged by forcing the ring down over a truncated conical block which is placed on the anvil for the purpose, and subsequently another heat is taken, and the hammering continued on the inclined beck of an anvil, as already described. The weight of the block can be accurately adjusted by varying the thickness at the time of punching out the central disk, by which means the amount of metal removed will be effected. Another plan adopted by Mr. Allen is to cast annular ingots, sometimes a number, one above the other, fed from one gate. These are cast with considerable depth, so as to allow of sufficient hammering to thoroughly consolidate the metal, and the weight is regulated by the size of the central core employed. For rolling the tires from the hammered rings he employs the tire-mill, constructed by Messrs. Galloway & Sons, of Manchester, which is the simplest one in use, and gives results probably not at all inferior to those of other more complicated forms. It is the one most generally adopted in England. The only other variation in the tire-making process is, that at some works, for the purpose of avoiding the severe one-sided strain brought upon the hammer by the use of the inclined beck for bevelling the rings, the ring is placed on a stout mandrel supported on a bifurcated anvil, and the necessary bevel is given by a tool of the proper shape with which the hammer is furnished. In Galloway's and most other tire-rolling machines the roll spindles are placed vertically and extend to a considerable distance below the horizontal bed of the machine. The rolls themselves are situated just above the surface of the latter, with no bearing above them, the spindles being long and stiff enough to resist all the strain coming upon them. The tire is thus readily dropped over the ends of the rolls and removed when finished. Its diameter is determined by a simple sliding gauge, measuring from the centre of the internal roll to the inner face of the tire at its greatest distance from the former. Bessemer steel tires by the above processes are now made in great numbers and give good satisfaction in use. There are some who still prefer the crucible steel for this purpose, but the difference in cost is so largely in favor of the Bessemer metal that it is probable the former will eventually cease to be made.

MANUFACTURE OF BESSEMER PLATES.

The application of the Bessemer process to the production of plates either for boilers or for ships, girders, etc., is one of the most important that could be made. Nevertheless the amount of metal used for this purpose in England falls much below that employed for other purposes. This is due to a certain amount of distrust of steel plates, doubt as to its reliability under varying strains of tension and compression, its capability of being punched and sheared without injury to itself, and of its action under the influence of heat and water, as in the fire-box of a boiler. In other countries, as for example Austria, as will be shown when we come to speak of the manufacture as carried on in that country, this has not been the case, and large quantities of plates have been produced and successfully applied to a variety of uses.

The secret of the distrust in regard to Bessemer plates in England is that in nearly all cases the percentage of carbon contained in the metal has been too large. The spiegeleisen used in England is not particularly rich in manganese—seldom exceeding nine per cent. of that element, while it generally contains from four to four and a half per cent. of carbon. It is difficult, therefore, with such materials to deoxygenate the metal sufficiently without introducing also a considerable percentage of carbon. About 0.4 per cent. of the latter is as large an amount as is proper for plates which are to resist severe strains, and though a greater proportion adds materially to the tensile strength of the metal when measured simply by a direct pull, it renders it also much harder and more liable to crack under the treatment to which it is exposed in the ordinary methods of construction. The difficulty in the way of producing good soft plates for boilers or other uses appeared at one time to have been satisfactorily overcome by the substitution of ferro-manganese in the place of the ordinary spiegeleisen. The manufacture of this substance was commenced by a firm in Glasgow as a branch of another business in which they were engaged, and plates made with it as a deoxygenator gave most excellent results. Unfortunately, however, the firm who had undertaken the manufacture shortly afterward became insolvent, and the patentee of the process has not as yet re-established the manufacture (which requires a considerable expenditure for suitable furnaces) elsewhere in England. Had the use of this substance continued for a longer time, so as to make the excellence of the steel produced with it fully appreciated by the public, there would have been a demand for plates urgent enough to have immediately secured the re-establishment of the manufacture; but in the present state of feeling it may not be so easy to induce the necessary primary outlay, especially as a certain amount of ill feeling is said to exist between the owners of the ferro-manganese patent and the Bessemer interest. The percentage of manganese contained in the alloy produced by the process referred to varied from fifteen to twenty-five. Another kind of ferro-manganese, containing a much larger percentage, and produced in Germany by a different process, also the subject of a patent, has been offered in the English market, but at such an exorbitant price that nobody has ventured to buy it. Still, notwithstanding the absence of ferro-manganese, good soft plates are produced at some works, especially those at Bolton. Messrs. Chas. Cammell & Co. also make a large number of plates of good quality. The following tests, which they guarantee all their plates to stand, are interesting:—

Tensile strain per square inch—thirty-three tons:—

Forge test (hot).—All plates one inch thick and under

to bend hot without fracture to an angle of 180°, both lengthways of the grain and across.

Forge test (cold).—All plates will admit of bending cold without fracture as follows:—

BESSEMER PLATES.

	With the grain.	Across the grain.
1 inch	45°	25°
$\frac{7}{8}$ inch	50	30
$\frac{3}{4}$ inch	60	40
$\frac{5}{8}$ inch	70	50
$\frac{1}{2}$ inch	80	60
$\frac{7}{16}$ inch	90	70
$\frac{3}{8}$ inch	110	80
$\frac{5}{16}$ inch	120	90
$\frac{1}{4}$ inch and upwards . .	120	100

To show the comparison of this steel with the regular crucible steel, the guarantee for plates of the latter is also given.

CRUCIBLE STEEL PLATES.

Tensile strain per square inch—thirty-eight tons.

	With the grain.	Across the grain.
1 inch	50°	30°
$\frac{7}{8}$ inch	60	35
$\frac{3}{4}$ inch	75	50
$\frac{5}{8}$ inch	90	70
$\frac{1}{2}$ inch	110	90
$\frac{7}{16}$ inch	130	100
$\frac{3}{8}$ inch	150	110
$\frac{5}{16}$ inch	180	120
$\frac{1}{4}$ inch and upwards . .	180	120

Probably the spiegeleisen used for this purpose is selected with especial care, and may contain as much as eleven per cent. of manganese without an increased proportion of carbon. By a proper system of testing the ingots, as described above, there should be and is no difficulty in ascertaining just what percentage of carbon is contained in the metal, and so selecting ingots that are suitable for this purpose. With the superior franklinite that we possess, together with the purer irons, there is, apparently, no reason why we should not produce most excellent plates in large quantities, as is already done in Austria.

The manufacture of axles is carried on to a considerable extent, both for locomotives and railway carriages. Locomotive crank shafts are now more frequently made of this material than any other, and with a far greater exemption from breakages. These are usually forged from large rectangular ingots, and twisted to the proper angle as in the case of iron. To bring these large masses down properly with economy requires very heavy hammers, and to meet this want Mr. Ramsbottom has erected at Crewe a thirty-ton hammer, on his patent duplex principle. In order to dispense with the costly foundations necessary to sustain the impact of the falling tup in large hammers, Mr. Ramsbottom designed about five years since, a hammer in which the blow should be struck by two heavy masses mounted on wheels, and moving horizontally in opposite directions, so that their momentum should be annihilated in striking the ingot placed between them. In the first of these hammers, in which the weight of each tup was ten tons, the cylinder was placed vertically in a pit beneath the hammer and the piston, connected by inclined links to each tup, so as to communicate motion to them on the rails. The ingot was supported on a suitable table, or between a pair of stout centres, which again rested on a platform capable of being rocked slightly to maintain the ingot always exactly in the centre of the motion of the tups. A number of these hammers are at present in use, and though they constitute the first development of a new idea, they do their work tolerably well, though they need a greater amount of care than an ordinary hammer. In the thirty-ton hammer which has been more recently built, the design has been somewhat modified, and greater simplicity obtained. In this the steam cylinders are horizontal, and placed directly behind each tup, the piston rods being secured to the latter by an elastic packing, so as to relieve the piston from the shock of the blow. To control the motion of the two tups, so that they shall always meet at the same point, a five-threaded screw with a diameter of six inches and a nine-inch pitch, or once and a half its diameter, is placed beneath them, the thread being cut left handed at one end, and right handed at the other. A nut secured to the bottom of each tup works on the portion of the screw beneath it, and as the screw revolves in its bearings each tup advances by the same amount. This arrangement is found to work with but little friction, and is not liable to derangement. The valve gear is made to be worked by hand in the ordinary way. The size of the cylinders and pressure of steam are so proportioned as to make the pressure on each tup the same as its weight, and the blow struck by this hammer is therefore the same as would be given by one of the tups falling by gravity through a distance equal to the combined stroke of the two tups, or seven feet. These hammers have been constructed by Messrs. Thwaites & Carbutt, of Bradford, who have had great experience in this line of business, having perhaps supplied more hammers to the steel makers than any other firm. With the heavy hammers just described, the large ingots for crank axles are brought down to the required size and shape in a very short time. At Crewe it is usual to put two of these ingots into the Siemens furnaces in the evening, and allow them to heat slowly during the night, but one man being required to be in attendance, and then to work them off under the hammer in the morning before breakfast. In sawing off the ends of his finished axle forgings, Mr. Ramsbottom employs a saw seven feet six inches in diameter, running at about nine hundred revolutions per minute, or a speed on the edge of four miles per minute. The cheeks are also sawed out preparatory to turning the crank wrists.

In concluding the account of the Bessemer manufacture, as at present conducted in England, we may observe that while the amount produced is far in excess of that to be found elsewhere, yet from the close competition between the different makers tending to favor the use of the cheapest materials, and from the naturally rather inferior character of the native iron employed, the quality of the metal is not equal to that produced in countries using better materials. Accordingly the uses to which it has been chiefly devoted have been rails, tires, and axles, together with a certain amount of plates. Notwithstanding this there have been produced, when proper substances have been employed, specimens of the metal which seemed able to undergo almost any test that could be devised. It has been spun into ornamental vessels of shapes such as would bring the most severe strain on the metal without exhibiting any sign of cracking, or bent into the most crucial shapes, with equal evidence of its toughness. We shall see on examining the product of other countries that such qualities in the metal are not at all exceptional, but that when steel of great hardness is not intentionally produced, they always exist.

SWEDEN.

An examination of the specimens of Bessemer steel from Sweden in the Exposition shows us that the metal there produced is of a far superior character to that made in England, and naturally leads to inquiry as to the cause of the difference, and whether we may hope to attain the same success in the United States. First we observe coils of wire of all sizes, down to the very finest, such as No. 47, or even smaller. This they have not been able regularly to produce in England. In the next place we notice a good display of fine cutlery, and the writer is informed by a competent authority that this metal answers so well for this purpose that it is now used almost to the exclusion of any other. This statement is corroborated by the fact that in the miscellaneous classes of the Swedish department, where cutlery occurs not as an exhibition of steel, but merely as a display of workmanship by other parties in the same manner as other articles of merchandise, cases of razors are exhibited with the mark of the kind of steel of which they are made stamped or etched upon them as usual, and these are all "Bessemer," but from a variety of different works, viz , Högbo, Carlsdal, Österby & Söderfors. The ore used in Sweden for producing iron for the Bessemer process is exclusively magnetic, and of a very pure quality. An analysis of a mixture of those used for the iron employed at the Fagersta works before roasting gives the following composition:—

Carb. acid	8.00
Silicium	17.35
Alumina	0.95
Lime	6.50
Magnesia	4.35
Protoxide of manganese	3.35
Magnetic oxide	32.15
Peroxide of iron	27.40
	100.05
Phosphoric acid	.03

All the pig made from this mixture of ores the exhibitors state will give a steel without the use of spiegeleisen, which is not at all red short.

The analysis of gray iron from the same works, used for the Bessemer process, is given as follows:—

Carbon combined	1.012
Graphite	3.527
Silicium	0.854
Manganese	1.919
Phosphorus	0.031
Sulphur	0.010

The cinder, produced at the same time as the gray iron, shows on analysis a composition of—

Silica	53.30
Alumina	3.00
Lime	21.10
Magnesia	13.95
Protoxide of manganese	7.85
Protoxide of iron	0.90
	100.10

The analysis of mottled pig (*fonte truitée*) consisting of two-thirds gray and one-third white, is—

Carbon combined	2.138
Graphite	2.733
Silicium	0.641
Manganese	2.926
Phosphorus	0.026
Sulphur	0.015

Of each of these it is stated that the steel produced without the employment of spiegeleisen is not at all red short (*cassant à chaud*). The most noticeable feature in the composition of these irons is the large percentage of manganese which they contain, together with the extremely minute proportion of sulphur. The latter quality is due to the exclusive employment of charcoal in the blast furnaces, together with the adoption of a very high temperature in the roasting kiln. These latter are constructed on Westman's patent, and are made very high and heated by the waste gas drawn from the blast furnaces. The heat is carried as high as is possible without agglomerating the materials, and by this treatment the ore is changed from a hard and compact substance to a very porous one, while at the same time it is stated that any percentage of sulphur less than four per cent. is driven off. The blast furnaces are very small, being generally but eight feet in diameter at the boshes and about three feet at the hearth, with a height of forty feet. With these ores prepared in this manner, such a furnace will yield from seventy to eighty tons per week. It is thought by the best informed engineers in Sweden that these furnaces should be made larger, and in future they probably will be so; but these dimensions represent the furnaces that now exist, and with which the iron in use has been produced.

In the process of conversion, from motives of economy, a fixed form of vessel is employed, instead of one mounted on trunnions, as in England and elsewhere. The tuyeres, about nineteen in number, are placed horizontally just above the bottom of the vessel, and are inclined a little from a radial direction so as to give a rotary motion to the mass of molten metal. An air passage surrounds the vessel at the back of the tuyeres, with a movable plate opposite each to allow access to them. The upper portion of the vessel, from the line of the top of the blast passage, is made removable, for lining, etc.; the bottom of the vessel is slightly inclined towards the taphole, so that the whole of the metal and slag may run off. The metal is run in at a spout in the upper portion of the vessel, and from the fixed position of the vessel it is of course necessary to have the blast on all the time that the metal is being run in and drawn off, to prevent its flowing into the tuyeres. This fact must make it more difficult to regulate the exact amount of decarbonization of the metal, and tend to render the last portion drawn off overdone. The removal of the cinder remaining in the vessel after a blow is not so easily accomplished in the fixed vessel as in the revolving one, as ordinarily used.

Accompanying the analyses of ores and irons, given above, the Fagersta works exhibit an analysis of the slag from the converter, taken at the close of the process, and it shows the composition to be as follows:—

Silica	44.30
Alumina	10.85
Lime	0.65
Magnesia	0.45
Protoxide of manganese	24.55
Protoxide of iron	19.45
	100.25

The case of specimens exhibited by these works is the most interesting by far in the Exposition. It contains a most extensive collection of pieces of various forms, with which a very elaborate set of experiments has just been made at Mr. D. Kirkaldy's testing works at London, the results of which will be found in Appendix C. The samples are classified according to the percentage of carbon which they contain, and have been tested to show their action under strains of tension, compression, torsion, bending, and, in the case of plates, bulging.

The amount of carbon contained in the steel varies from 0.1 to 1.50 per cent., though most of the experiments were made between the limits of 0.3 and 1.20 per cent. In addition to the large collection of test pieces, they exhibit some railway carriage axles containing 0.3 per cent. of carbon, one being bent double with a radius of curvature at the bend of about five inches; a locomotive axle containing 0.4 per cent., and a tire having 0.5 per cent. of carbon. There is also, as already mentioned, a fine display of cutlery, razors, some beautiful hand mirrors containing 1.0 per cent., a small drill containing 1.50 per cent., with a plate beside it containing 1.00 per cent., through which it had drilled several holes; a number of long turnings taken off in a lathe, showing remarkably the absolute continuity of the grain—one of 0.3 per cent. of carbon measures 36 feet in length, and is closely coiled with a diameter of about $\frac{1}{12}$ inch; another of 0.9 per cent. is 27 feet long and slightly less in diameter. There are also a large number of files, and, as previously mentioned, coils of wire of all sizes, and apparently any required length. A very interesting table of results was obtained from a series of eleven small square bars containing varying percentages of carbon, as follows:—

No.	Per cent. of carbon.	Sectional area before elongation, square inches.	Breaking weight, in pounds.	Breaking weight, per square inch.	Section after fracture at point of rupture.	Proportion of ruptured section to original section.	Breaking weight per square inch of ruptured section.	Per cent. of elongation.
1	0.35	.2323	16,262	69,730	.0854	36.65	190,250	12.0
2	0.45	.1448	14,663	100,800	.0996	68.5	147,160	10.3
3	0.45	.1398	14,663	104,300	.1150	81.9	130,300	9.2
4	0.70	.234	29,540	125,800	.2026	86.3	145,750	1.56
5	0.70	.1568	16,074	102,300	.1314	83.46	122,300	4.0
6	0.70	.1515	19,841	131,400	.1400	92.05	141,660	5.4
7	0.70	.1485	17,016	114,100	.1230	82.55	138,240	5.8
8	0.90	.1466	19,935	135,400	.1189	80.80	167,500	6 7
9	1.00	.2338	30,012	128,000	.2242	95.69	133,800	2.3
10	1.00	.1516	20,218	132,700	.1400	91.93	144,300	6.6
11	1.00	.1494	21,726	144,800	.1400	93.31	155,120	4.0

The cost of steel for the more delicate uses, such as razors, etc., is very much less by the Bessemer process than by the old method of remelting in the crucible. The materials in ordinary use are sufficiently pure to give such a steel, and the only special precaution which has to be observed in producing these qualities is to add a sufficient amount of recarbonizing pig to give the required per cent. of carbon, and then in the process of tilting the bars to carefully reject any piece which may show sign of flaw, as would of course be necessary under any circumstances. The total production of Bessemer steel in Sweden in 1864 was 3178 tons; that of crucible steel exceeded 4500 tons.

AUSTRIA.

The conditions under which Bessemer metal is produced in Austria are in many respects similar to those existing in Sweden. The iron employed is smelted with charcoal, is nearly free from sulphur and phosphorus, and contains a large percentage of manganese. There are differences in the manner of conducting the process, but these important conditions insure the production of a metal of similar excellence to the Swedish, and, like this, much superior to the ordinary metal produced in England.

The principal works in Austria are at Neuberg, in the province of Styria, and are carried on by the government. The iron is obtained from spathic ores smelted in two furnaces 43 feet high, and yielding from 100 to 150 tons per week. The iron produced is found by analysis to contain 3.46 per cent. of manganese, and, as in Sweden, it is used for recarbonizing in the place of the usual spiegeleisen. Originally a fixed vessel was erected as these works similar to those used in Sweden, but this has been superseded by a pair of three-ton vessels of the ordinary construction. Fixed or Swedish vessels are, however, still in use at other Austrian works. The metal it run directly from the blast furnaces into the converters. Very interesting tables are exhibited by these works, giving analyses of the iron and slag at five periods in its conversion from its condition as tapped from the furnace to its final state as Bessemer metal. These are extremely interesting from the light which they throw upon the relative rapidity with which the components of the pig iron are attacked by the blast, and the permanency of some ingredients, such as phosphorus and copper, during the entire process. The results are as follows:—

	As tapped from blast furnace.	After the disappearance of the sparks from the converter.	After the boiling over period.	End of blowing.	After addition of pig for recarbonization.
IRON.					
Graphite	3.180	....	....	....	....
Carbon combined	0.750	2.465	0.949	0.087	0.234
Silicium	1.960	0.443	0.112	0.028	0.033
Phosphorus	0.040	0.040	0.045	0.045	0.044
Sulphur	0.018	trace	trace	trace	trace
Manganese	3.460	1.645	0.429	0.113	0.139
Copper	0.085	0.091	0.095	0.120	0.105
Iron	90.507	95.316	98.370	99.607	99.445
SLAG.					
Silica	40.95	46.78	51.75	46.75	47.25
Alumina	8.70	4.65	2.98	2.80	3.45
Protoxide of iron	0.60	6.78	5.80	16.86	15.43
Protoxide of manganese	2.18	37.00	37.90	32.23	31.89
Lime	30.35	2.98	1.76	1.19	1.23
Magnesia	16.32	1.53	0.45	0.52	0.61
Potash	0.18	trace	trace	trace	trace
Soda	0.14	trace	trace	trace	trace
Sulphur	0.34	trace	trace	trace	trace
Phosphorus	0.01	0.03	0.02	0.01	0.01

From each charge blown at these works a small test ingot is cast, and this is immediately reheated and subjected to a number of tests to ascertain the quality of the steel; and according to the results of these trials, all the metal produced is divided into seven grades of varying hardness, No. 1 being a blue steel, containing from 1.12 to 1.58 per cent. of carbon; and No. 7 a soft iron, with from 0.05 to 0.15 per cent.

The test employed consists in hammering the little

ingot into a bar, and subjecting it to severe working on the anvil, in a way which would tend to crack it if of a red short nature, or of inferior quality. It is then heated and plunged into water, and the amount of hardening produced proved by striking it with a hammer, and observing the amount of flexure produced. It is then heated again and bent over upon itself and welded into an eye, the welded portion being drawn out to a small section and broken off. These tests take but a short time, and the expense of making them is insignificant in comparison with the accurate knowledge thereby obtained of the nature of the steel and the purposes for which it is suitable. As a rule, the steel produced at the Neuberg works welds with great facility, and, in fact, all the tires produced here are welded as in the case of iron. A table of the tensile strengths and other properties of steel, of the various classes below No. 2, is exhibited, and is as follows :—

	No. 3.	No. 4.	No. 5.	No. 6.	No. 7.
Percentage of combined carbon.	0.88 to 1.12	0.62 to 0.88	0 38 to 0.62	0.15 to 0.38	0.05 to 0.15
Tensile strength, tons per square inch.	63.13 to 74.61	51.65 to 63.13	40.17 to 51.65	34.43 to 40.17	28.69 to 34.43
Extensibility . .	.05	.10 to .05	.20 to .10	.25 to .20	.30 to .25
Hardening . . .	with care	very well	very well	feebly	not at all
Welding . . .	very well as hard cast steel	very well	very well	very well	very well

The softest grade is used for wire, sheet steel, etc., and the higher numbers for boiler plate, gun barrels, axles, tires, tools, and cutlery, according to the hardness required.

A printed list gives the price of the steel in various forms delivered at the works, which, reduced to gold dollars, is as follows: ingots, $77.50; bars, $138; boiler plate, $145.50; tires, $155.50. These prices are little above those charged in England, where coal is abundant and an inferior quality of metal produced.

In other countries than Sweden and Austria, we find nothing that presents any remarkable feature not to be found in English practice. Of course, Krupp is far ahead of all others in respect to the size of the masses that he casts. He exhibits in the Exposition a forty-ton (40,000 kilograms) ingot, intended for a crank shaft, which he states was cast from crucibles. His process of making tires is similar to that in use in England. He first makes a bloom about 6 feet long and 13 inches by 10 inches, and then cuts this up into sections of the required weight. A slit is cut through the middle of these, and they are then worked out into an annular form, and afterwards rolled on a mill of a construction similar to those in use in England, with the exception that the bed, instead of being horizontal, is vertical, as if one of those machines were turned up on its edge. Two mills, one for roughing and one for finishing, are employed. His tire-heating furnaces are placed in a pit at the side of the mill, and are similar to the furnaces of a brass foundry, the tires being laid on the fire by a central crane.

The French also exhibit good specimens of Bessemer metal, but, as already stated, there seems to be no marked advance on what has been accomplished in England, and it will not be necessary, therefore, to notice in detail the articles they have brought forward.

The manufacture has been established at six works, and the production, in 1866 was as follows:—

	Tons.
Compagnie de Terrenoire	1,537
Cie. de Chatillon, Commentry	59
Société d'Imphy, St. Seurin (Jackson's)	4,858
S. Menan's & Cie	000
De Dietrich & Cie	486
Petin, Gaudet & Cie	3,851
Total	10,791

Of this product, 3687 tons were in the form of rails. In 1863 but three works were in operation, with a total product of 1857 tons. At the present time the metal produced in France by this process does not stand as high in the opinion of iron-masters as puddled or other steel. It may be that this is due to the nature of the pig iron employed, or it may be due to a lack of experience in the manufacture as compared with other nations.

At the works of Messrs. Petin, Gaudet & Co., near St. Etienne, a pair of six-ton converters have been erected, and a single vessel, capable at present of producing a charge of eight tons, and in which it is expected to make twelve-ton charges when the lining becomes reduced in thickness. This is the largest Bessemer apparatus in France.

Submitted by FREDERIC J. SLADE,
Scientific Assistant to Committee No. 6.

PARIS, June 15, 1867.

Berard and Martin Processes.

A careful study of the Exposition showed but two other processes for making steel worthy of notice, and both French: the one patented by A. Berard and tried at the forges of Montataire; the other that of Emille and Pierre E. Martin, in operation at Sireuil. In both these systems cast steel is made in a reverberatory furnace. In Berard's process the conversion of the pig iron into steel is sought to be achieved by subjecting the melted metal alternately to a decarbonizing and recarbonizing flame, for which purpose it is necessary to employ blast. He uses a Siemens furnace, and avails himself of the changes of current required in working the regenerators to effect the changes of flame. The furnace is divided by a bridge into two halves, and he thus operates upon two masses of iron at the same time, one of which is freshly charged, while the other contains material which is nearly decarbonized. Some specimens of Berard's steel were on exhibition, and although creditable in themselves, it was generally understood that he had not yet succeeded in making steel regularly for market. The Messrs. Martin, on the contrary, were not only making steel regularly at their own works at Sireuil, but the process is also in operation at two of the largest works in France—Le Creusot and Firminy, and is in process of erection at various other works in Europe, and arrangements have been made for its immediate introduction into the United States. In this process the pig iron is deprived of its carbon by the addition of pieces of wrought iron or steel either in the form of shingled puddle balls, or of scrap. The quantity, however, of wrought iron necessary to reduce the carbon to the required limits, is much less than would be inferred, from the consideration of the quantity contained in the pig, and does not in practice much exceed the quantity of pig itself. A charge of gray pig or of spiegeleisen is melted in a Siemens furnace, having a bed hollowed out to contain it, and is allowed to remain about half an hour after fusion to bring it to an intense white heat; portions of malleable iron previously brought to a bright red heat are then added in successive charges of about 200 pounds, at intervals of twenty minutes to a half hour, each charge being thoroughly melted before the next is added. After two or three such additions, ebullition commences in the bath of metal, and continues till the carbon is wholly removed from the pig. The exact condition of the metal is ascertained from small proofs taken from the charge, after each addition of iron towards the end of the operation. These are run into a small ingot mould, and when cooled to the proper heat, hammered into a plate, about $\frac{5}{16}$ of an inch thick by 5 inches in diameter. When the decarbonization is completely effected these proofs will bend double cold, and show a fracture quite fibrous. A quantity of pig, generally of the same kind as was used for the preliminary charge, is then added in such proportion to the amount of iron in the furnace as to give the desired hardness to the steel, according to the use for which it is required. When this is melted the bath is well stirred to insure homogeneity in its substance, and a final proof taken, which is treated in the same manner as the others, and gives reliable evidence as to the state of the metal before pouring. This enables the quality to be very exactly adjusted to the degree of hardness required. Should it be too soft, more pig is added, while if it is too hard, the mere waiting from a quarter to half an hour will materially soften the metal. Arguing from this fact, Messrs. Martin claim that under the influence of such a high temperature, the carbon is to some extent spontaneously disassociated from the iron, and attribute in a measure to this fact that so small a proportion of wrought iron is required to effect the decarbonization of the pig. The coating of scale formed on the iron in the preliminary reheating which it undergoes before being charged into the furnace, also assists in the removal of the carbon. When the metal has been brought to the desired condition, it is tapped off at the rear of the furnace into ingot moulds placed on a railway car, and thus brought successively under the gutter.

A considerable number of specimens of steel made by this process were exhibited, ranging in hardness from a metal too hard to be touched by a tool to a true wrought iron, intended to be used in the manufacture of armor plates. At Messrs. Martins' works, at Sireuil, the process has been in regular operation during the past two years for the manufacture of gun-barrels, and some remarkable specimens of these were exhibited. Thus there was one that had been tested with very large charges of powder and a heavy weight of shot, which, by very palpable bulging just behind the balls, testified as to the softness and toughness of the metal. In another, which had been burst by a similarly severe charge, the metal had merely torn open for a certain length of the barrel, and the lips so formed were simply folded back 180'', without any sign of cracking. There were also shown specimens of tool-steel of excellent fracture, castings of pieces of machinery, such as gears and framing, and a large tube for a cannon of extremely soft metal, or melted iron, as it is named.

The hardest variety of metal, called by the patentee "mixed metal," is considered suitable for castings which do not require to be worked by tools, but where great strength is required, such as hammer blocks and anvils, large gears, etc. By a subsequent process of annealing or discarbonization, carried on in a gas furnace, under the influence of an oxidizing flame, these castings may be softened so as to be quite malleable and easily worked, and they then retain the advantage of being free from blow-holes. This metal is produced by adding to a preliminary bath of say 1600 pounds of pig 2400 of wrought iron, and adding at the end 1200 pounds of pig. For tool-steel, to a bath of 1600 pounds of gray pig would be added 2600 pounds of puddled steel from the same pig, and at the end of the operation 400 to 500 pounds of spiegeleisen. For homogeneous metal, the preliminary bath at Sireuil is 1200 pounds of spiegeleisen, to which 2000 pounds of soft iron, puddled to grain, from the same pig, is added, and at the end of the process 200 to 300 pounds of the same pig is charged, to give the requisite amount of carbon. The softest metal of all, which, however, has not as yet been made an article of regular manufacture, is made in the same way, with the exception that the final charge of manganiferous pig is but 5 per cent. of the contents of the furnace. With certain kinds of gray charcoal pig this proportion rises, however, to 20 per cent., since under the influence of the high temperature they refine spontaneously with great rapidity.

Messrs. Martins' patents also cover the use of ore either with or in place of the wrought iron or steel used for removing the carbon from the pig, and when this is used the progress of the operation is much more rapid. It has the objection, however, that the slag formed attacks violently the bricks forming the sides of the furnace, and therefore requires frequent renewals.

This process has the great practical advantage that all the scrap arising in the manufacture of any product, such as the ends of bars, etc., is readily remitted in the furnace and immediately returned to the form of useful ingots.

The flame in the furnace is kept always slightly surcharged with gas; an effect which the use of the Siemens furnace renders easy and certain, and by this means the waste of the metal is always moderate.

For the production of soft steel suitable for gun-barrels or for tires, this metal already enjoys considerable reputation in Europe, and, indeed, were it not for its excellent quality, it would be impossible to sustain the manufacture at Sireuil, where there is neither iron nor coal, the latter being brought from England and the former from various parts of France.

The results here stated were verified by a personal residence of Mr. Slade during several weeks at the works at Sireuil, and the regular and commercial success of the process was in that way seen to be fully achieved.

It is not asserted that cast-steel can be made as cheaply by this process as by the Bessemer; but where a product of definite quality is to be produced day by day, without rejections to any considerable extent, the Martin process has a decided advantage over the Bessemer, and in comparison with the crucible steel is decidedly less expensive. Its chief drawback would seem to lie in the difficulty of keeping the furnace in order, and only the most refractory materials will withstand the high heat required for its operation. As much as five tons of steel have been produced by this process at a single heat, and there is no difficulty in combining the product of several furnaces where larger masses are desired, inasmuch as the temper of the heat in each furnace can be brought and maintained to exactly the same standard. It would seem also to present the best solution yet devised for the difficulty experienced by the accumulation of the ends of Bessemer steel rails, inasmuch as these can be used in lieu of the puddled iron required by the process. It is possible, also, to use old rails in the same manner, and, indeed, any old scrap, but the resulting quality of the steel will, to a great extent, depend upon the quality of the old iron so used.

TABLES

SHOWING THE

RELATIVE VALUES OF FRENCH AND ENGLISH WEIGHTS AND MEASURES, &c.

Measures of Length.

Millimetre	=	0.03937	inch.
Centimetre	=	0.393708	"
Decimetre	=	3.937079	inches.
Metre	=	39.37079	"
"	=	3.2808992	feet.
"	=	1.093633	yard.
Decametre	=	32.808992	feet.
Hectometre	=	328.08992	"
Kilometre	=	3280.8992	"
"	=	1093.633	yards.
Myriametre	=	10936.33	"
"	=	6.2138	miles.
Inch ($\frac{1}{36}$ yard)	=	2.539954	centimetres.
Foot ($\frac{1}{3}$ yard)	=	3.0479449	decimetres.
Yard	=	0.91438348	metre.
Fathom (2 yards)	=	1.82876696	"
Pole or perch ($5\frac{1}{2}$ yards)	=	5.029109	metres.
Furlong (220 yards)	=	201.16437	"
Mile (1760 yards)	=	1609.3149	"
Nautical mile	=	1852	"

Superficial Measures.

Square millimetre	=	$\frac{1}{645}$	square	inch.
" "	=	0.00155	"	"
" centimetre	=	0.155006	"	"
" decimetre	=	15.50059	"	inches.
" "	=	0.107643	"	foot.
" metre or centiare	=	1550.05989	"	inches.
" " "	=	10.764299	"	feet.
" " "	=	1.196033	"	yard
Are	=	1076.4299	"	feet.
"	=	119.6033	"	yards.
"	=	0.098845	rood.	
Hectare	=	11960.3326	square	yards.
"	=	2.471143	acres.	
Square inch	=	645.109201	square	millimetres.
" "	=	6.451367	"	centimetres
" foot	=	9.289968	"	decimetres.
" yard	=	0.836097	"	metre.
" rod or perch	=	25.291939	"	metres.
Rood (1210 sq. yards)	=	10.116775	ares.	
Acre (4840 sq. yards)	=	0.404671	hectare.	

Measures of Capacity.

Cubic millimetre	=	0.000061027	cubic	inch.
" centimetre or millilitre	=	0.061027	"	"
10 " centimetres or centilitre	=	0.61027	"	"
100 " " " decilitre	=	6.102705	"	inches.
1000 " " " litre	=	61.0270515	"	"
" " " " "	=	1.760773	imp'l	pint.
" " " " "	=	0.2200967	"	gal'n.
Decalitre	=	610.270515	cubic	inches.
"	=	2.2009668	imp.	gal'ns.
Hectolitre	=	3.531658	cubic	feet.
"	=	22.009668	imp.	gal'ns.
Cubic metre or stere or kilolitre	=	1.30802	cubic	yard.
" " "	=	35.3165807	"	feet.
Myrialitre	=	353.165807	"	"

Cubic inch	= 16.386176	cubic centimetres.
" foot	= 28.315312	" decimetres.
" yard	= 0.764513422	" metre.

American Measures.

Winchester or U.S. gallon (231 cub.in.)	= 3.785209	litres.
" " bushel (2150.42 cub. in.)	= 35.23719	"
Chaldron (57.25 cubic feet)	= 1621.085	"

British Imperial Measures.

Gill	= 0.141983	litre.
Pint ($\frac{1}{8}$ gallon)	= 0.567932	"
Quart ($\frac{1}{4}$ gallon)	= 1.135864	"
Imperial gallon (277.2738 cub. in.)	= 4.54345797	litres.
Peck (2 gallons)	= 9.0869159	"
Bushel (8 gallons)	= 36.347664	"
Sack (3 bushels)	= 1.09043	hectolitre.
Quarter (8 bushels)	= 2.907813	hectolitres.
Chaldron (12 sacks)	= 13.08516	"

Weights.

Milligramme	=	0.015438395	troy grain.
Centigramme	=	0.15438395	" "
Decigramme	=	1.5438395	" "
Gramme	=	15.438395	" grains.
"	=	0.643	pennyweight.
"	=	0.0321633	oz. troy.
"	=	0.0352889	oz. avoirdupois.
Decagramme	=	154.38395	troy grains.
"	=	5.64	drachms avoirdupois.
Hectogramme	=	3.21633	oz. troy.
"	=	3.52889	oz. avoirdupois.
Kilogramme	=	2.6803	lbs. troy.
"	=	2.205486	lbs. avoirdupois.
Myriagramme	=	26.803	lbs. troy.
"	=	22.05486	lbs. avoirdupois.

Quintal metrique = 100 kilog. = 220.5486 lbs. avoirdupois.
Tonne = 1000 kilog. = 2205.486 " "

29

Different authors give the following values for the gramme:—

Gramme	= 15.44402	troy grains.
"	= 15.44242	"
"	= 15.4402	"
"	= 15.433159	"
"	= 15.43234874	"

AVOIRDUPOIS.

Long ton = 20 cwt. = 2240 lbs.	= 1015.649	kilogrammes.
Short ton (2000 lbs.)	= 906.8296	"
Hundred weight (112 lbs.)	= 50.78245	"
Quarter (28 lbs.)	= 12.6956144	"
Pound = 16 oz. = 7000 grs.	= 453.4148	grammes.
Ounce = 16 dr'ms. = 437.5 grs.	= 28.3375	"
Drachm = 27.344 grains	= 1.77108	gramme.

TROY (PRECIOUS METALS).

Pound = 12 oz. = 5760 grs.	= 373.096	grammes.
Ounce = 20 dwt. = 480 grs.	= 31.0913	"
Pennyweight = 24 grs.	= 1.55457	gramme.
Grain	= 0.064773	"

APOTHECARIES' (PHARMACY).

Ounce = 8 drachms = 480 grs.	= 31.0913	gramme.
Drachm = 3 scruples = 60 grs.	= 3.8869	"
Scruple = 20 grs.	= 1.29546	gramme.

CARAT WEIGHT FOR DIAMONDS.

1 carat = 4 carat grains = 64 carat parts.

"	= 3.2	troy grains.
"	= 3.273	" "
"	= 0.207264	gramme
"	= 0.212	"
"	= 0.205	"

Great diversity in value.

Proposed Symbols for Abbreviations.

M—myria — 10000	Mm	Mg	Ml	
K—kilo — 1000	Km	Kg	Kl	
H—hecto — 100	Hm	Hg	Hl	Ha
D—deca — 10	Dm	Dg	Dl	Da
Unit — 1	metre—m	gramme—g	litre—l	are—a
d—deci — 0.1	dm	dg	dl	da
c—centi — 0.01	cm	cg	cl	ca
m—milli — 0.001	mm	mg	ml	

Km = Kilometre. Hl = Hectolitre. cg = centigramme. c. cm = $\overline{cm}^3$ = cubic centimetre. $\overline{dm}^2$ = sq. dm = square decimetre. Kgm = Kilogrammetre. Kg° = Kilogramme degree.

Celsius or Centigrade.	Fahrenheit.	Réaumur.
— 15°	+ 5°	— 12°
— 10	+ 14	— 8
— 5	+ 23	— 4
0 melting	+ 32	ice 0
+ 5	+ 41	+ 4
+ 10	+ 50	+ 8
+ 15	+ 59	+ 12
+ 20	+ 68	+ 16
+ 25	+ 77	+ 20
+ 30	+ 86	+ 24
+ 35	+ 95	+ 28
+ 40	+104	+ 32
+ 45	+113	+ 36
+ 50	+122	+ 40
+ 55	+131	+ 44
+ 60	+140	+ 48
+ 65	+149	+ 52
+ 70	+158	+ 56
+ 75	+167	+ 60
+ 80	+176	+ 64
+ 85	+185	+ 68
+ 90	+194	+ 72
+ 95	+203	+ 76
+100 boiling	+212	water + 80
+200	+392	+160
+300	+572	+240
+400	+752	+320
+500	+932	+400

1° C. = 1°.8 Ft. = $\frac{9}{5}$° Ft. = 0°.8 R. = $\frac{4}{5}$° R.

1° C. × $\frac{9}{5}$ = 1° Ft. 1° Ft. × $\frac{5}{9}$ = 1° C. 1° R. × $\frac{9}{4}$ = 1° Ft.

1° C. × $\frac{4}{5}$ = 1° R. 1° Ft. × $\frac{4}{9}$ = 1° R. 1° R. × $\frac{5}{4}$ = 1° C.

Calorie (French) = unit of heat } English.
= kilogramme degree }

It is the quantity of heat necessary to raise 1° C. the temperature of 1 kilogramme of distilled water.

Kilogrammetre = Kgm = the power necessary to raise 1 kilogramme, 1 metre high, in one second. It is equal to $\frac{1}{75}$ of a French horse power. An English horse power = 550 foot pounds, while a French horse power = 542.7 foot pounds.

Ready-made Calculations.

No. of units.	Inches to centimetres.	Feet to metres.	Yards to metres.	Miles to Kilometres.	Millimetres to inches.
1	2.53995	0.3047945	0.91438348	1.6093	0.03937079
2	5.0799	0.6095890	1.82876696	3.2186	0.07874158
3	7.6199	0.9143835	2.74315044	4.8279	0.11811237
4	10.1598	1.2197680	3.65753392	6.4373	0.15748316
5	12.6998	1.5239724	4.57191740	8.0466	0.19685395
6	15.2397	1.8287669	5.48630088	9.6559	0.23622474
7	17.7797	2.1335614	6.40068436	11.2652	0.27559553
8	20.3196	2.4383559	7.31506784	12.8745	0.31496632
9	22.8596	2.7431504	8.22945132	14.4838	0.35433711
10	25.3995	3.0479450	9.14383480	16.0930	0.39370790

No. of units.	Centimetres to inches.	Metres to feet.	Metres to yards.	Kilometres to miles.	Square inches to square centimetres.
1	0.3937079	3.2808992	1.093633	0.6213824	6.45136
2	0.7874158	6.5617984	2.187266	1.2427648	12.90272
3	1.1811237	9.8426976	3.280899	1.8641472	19.35408
4	1.5748316	13.1235968	4.374532	2.4855296	25.80544
5	1.9685395	16.4044960	5.468165	3.1069120	32.25680
6	2.3622474	19.6853952	6.561798	3.7282944	38.70816
7	2.7559553	22.9662944	7.655431	4.3496768	45.15952
8	3.1496632	26.2471936	8.749064	4.9710592	51.61088
9	3.5433711	29.5280928	9.842697	5.5924416	58.06224
10	3.9370790	32.8089920	10.936330	6.2138240	64.51360

No of units.	Square feet to sq. metres.	Sq. yards to sq. metres.	Acres to hectares.	Square centimetres to sq. inches.	Sq. metres to sq. feet.
1	0.0929	0.836097	0.404671	0.155	10.7643
2	0.1858	1.672194	0.809342	0.310	21.5286
3	0.2787	2.508291	1.204013	0.465	32.2929
4	0.3716	3.344388	1.618684	0.620	43.0572
5	0.4645	4.180485	2.023355	0.775	53.8215
6	0.5574	5.016582	2.428026	0.930	64.5858
7	0.6503	5.852679	2.832697	1.085	75.3501
8	0.7432	6.688776	3.237368	1.240	86.1144
9	0.8361	7.524873	3.642039	1.395	96.8787
10	0.9290	8.360970	4.046710	1.550	107.6430

No. of units.	Square metres to sq. yards.	Hectares to acres.	Cubic inches to cubic centimetres.	Cubic feet to cubic metres.	Cubic yards to cubic metres.
1	1.196033	2.471143	16.3855	0.02831	0.76451
2	2.392066	4.942286	32.7710	0.05662	1.52902
3	3.588099	7.413429	49.1565	0.08494	2.29354
4	4.784132	9.884572	65.5420	0.11325	3.05805
5	5.980165	12.355715	81.9275	0.14157	3.82257
6	7.176198	14.826858	98.3130	0.16988	4.58708
7	8.372231	17.298001	114.6985	0.19819	5.35159
8	9.568264	19.769144	131.0840	0.22651	6.11611
9	10.764297	22.240287	147.4695	0.25482	6.88062
10	11.960330	24.711430	163.8550	0.28315	7.64513

No. of units.	Cubic centimetres to cubic inches.	Litres to cubic inches.	Hectolitres to cubic feet.	Cubic metres to cubic feet.	Cubic metres to cubic yards.
1	0.06102	61.02705	3.5317	35.31659	1.30802
2	0.12205	122.05410	7.0634	70.63318	2.61604
3	0.18308	183.08115	10.5951	105.94977	3.92406
4	0.24411	244.10820	14.1268	141.26636	5.23208
5	0.30514	305.13525	17.6585	176.58295	6.54010
6	0.36617	366.16230	21.1902	211.89954	7.84812
7	0.42720	427.18935	24.7219	247.21613	9.15614
8	0.48823	488.21640	28.2536	282.53272	10.46416
9	0.54926	549.24345	31.7853	317.84931	11.77218
10	0.61027	610.27050	35.3166	353.16590	13.08020

No. of units.	Grains to grammes.	Ounces avoir. to grammes.	Ounces troy to grammes.	Pounds avoir. to kilogrammes.	Pounds troy to kilogrammes.
1	0.064773	28.3375	31.0913	0.4534148	0.373096
2	0.129546	56.6750	62.1826	0.9068296	0.746192
3	0.194319	85.0125	93.2739	1.3602444	1.119288
4	0.259092	113.3500	124.3652	1.8136592	1.492384
5	0.323865	141.6871	155.4565	2.2670740	1.865480
6	0.388638	170.0250	186.5478	2.7204888	2.238576
7	0.453411	198.3625	217.6391	3.1739036	2.611672
8	0.518184	226.7000	248.7304	3.6273184	2.984768
9	0.582957	255.0375	279.8217	4.0807332	3.357864
10	0.647730	283.3750	310.9130	4.5341480	3.730960

No. of units.	Long tons to tonnes of 1000 kilog.	Pounds per square inch to kilogrammes per square centimetre.	Grammes to grains.	Grammes to ounces avoir.	Grammes to ounces troy.
1	1.015649	0.0702774	15.438395	0.0352889	0.0321633
2	2.031298	0.1405548	30.876790	0.0705778	0.0643266
3	3.046947	0.2108322	46.315185	0.1058667	0.0964899
4	4.062596	0.2811096	61.753580	0.1411556	0.1286532
5	5.078245	0.3513870	77.191975	0.1764445	0.1608165
6	6.093894	0.4216644	92.630370	0.2117334	0.1929798
7	7.109543	0.4919418	108.068765	0.2470223	0.2251431
8	8.125192	0.5622192	123.507160	0.2823112	0.2573064
9	9.140841	0.6324966	138.945555	0.3176001	0.2894697
10	10.156490	0.7027740	154.383950	0.3528890	0.3216330

No. of units.	Kilogrammes to pounds avoirdupois.	Kilogrammes to pounds troy.	Metric tonnes of 1000 kilog. to long tons of 2240 pounds.	Kilog. per square millimetre to pounds per square inch.	Kilog. per square centimetre to pounds per square inch.
1	2.205486	2.6803	0.9845919	1422.52	14.22526
2	4.410972	5.3606	1.9691838	2845.05	28.45052
3	6.616458	8.0409	2.9537757	4267.57	42.67578
4	8.821944	10.7212	3.9383676	5690.10	56.90104
5	11.027430	13.4015	4.9229595	7112.63	71.12630
6	13.232916	16.0818	5.9075514	8535.15	85.35156
7	15.438402	18.7621	6.8921433	9957.68	99.57682
8	17.643888	21.4424	7.8767352	11380.20	113.80208
9	19.849374	24.1227	8.8613271	12802.73	128.02734
10	22.054860	26.8030	9.8459190	14225.26	142.25260

INDEX.

CATALOGUE

OF

PRACTICAL AND SCIENTIFIC BOOKS,

PUBLISHED BY

HENRY CAREY BAIRD,

INDUSTRIAL PUBLISHER,

No. 406 WALNUT STREET,

PHILADELPHIA.

☞ Any of the Books comprised in this Catalogue will be sent by mail, free of postage, at the publication price.

☞ This Catalogue will be sent, free of postage, to any one who will furnish the publisher with his address.

ARMENGAUD, AMOUROUX, AND JOHNSON.—THE PRACTICAL DRAUGHTSMAN'S BOOK OF INDUSTRIAL DESIGN, AND MACHINIST'S AND ENGINEER'S DRAWING COMPANION: Forming a complete course of Mechanical Engineering and Architectural Drawing. From the French of M. Armengaud the elder, Prof. of Design in the Conservatoire of Arts and Industry, Paris, and MM. Armengaud the younger and Amouroux, Civil Engineers. Rewritten and arranged, with additional matter and plates, selections from and examples of the most useful and generally employed mechanism of the day. By William Johnson, Assoc. Inst. C. E., Editor of "The Practical Mechanic's Journal." Illustrated by 50 folio steel plates and 50 wood-cuts. A new edition, 4to. $10 00

ARROWSMITH.—PAPER-HANGER'S COMPANION: A Treatise in which the Practical Operations of the Trade are Systematically laid down: with Copious Directions Preparatory to Papering; Preventives against the Effect of Damp on Walls; the Various Cements and Pastes adapted to the Several Purposes of the Trade; Observations and Directions for the Panelling and Ornamenting of Rooms, &c. By James Arrowsmith, Author of "Analysis of Drapery," &c. 12mo., cloth $1 25

BAIRD.—THE AMERICAN COTTON SPINNER, AND MANAGER'S AND CARDER'S GUIDE:

A Practical Treatise on Cotton Spinning; giving the Dimensions and Speed of Machinery, Draught and Twist Calculations, etc.; with notices of recent Improvements: together with Rules and Examples for making changes in the sizes and numbers of Roving and Yarn. Compiled from the papers of the late ROBERT H. BAIRD. 12mo. $1 50

BAKER.—LONG-SPAN RAILWAY BRIDGES:

Comprising Investigations of the Comparative Theoretical and Practical Advantages of the various Adopted or Proposed Type Systems of Construction; with numerous Formulæ and Tables. By B. Baker. 12mo. $2 00

BAKEWELL.—A MANUAL OF ELECTRICITY—PRACTICAL AND THEORETICAL:

By F. C. BAKEWELL, Inventor of the Copying Telegraph. Second Edition. Revised and enlarged. Illustrated by numerous engravings. 12mo. Cloth $2 00

BEANS.—A TREATISE ON RAILROAD CURVES AND THE LOCATION OF RAILROADS:

By E. W. BEANS, C. E. 12mo. (In press.)

BLENKARN.—PRACTICAL SPECIFICATIONS OF WORKS EXECUTED IN ARCHITECTURE, CIVIL AND MECHANICAL ENGINEERING, AND IN ROAD MAKING AND SEWERING:

To which are added a series of practically useful Agreements and Reports. By JOHN BLENKARN. Illustrated by fifteen large folding plates. 8vo. $9 00

BLINN.—A PRACTICAL WORKSHOP COMPANION FOR TIN, SHEET-IRON, AND COPPER-PLATE WORKERS:

Containing Rules for Describing various kinds of Patterns used by Tin, Sheet-iron, and Copper-plate Workers; Practical Geometry; Mensuration of Surfaces and Solids; Tables of the Weight of Metals, Lead Pipe, etc.; Tables of Areas and Circumferences of Circles; Japans, Varnishes, Lackers, Cements, Compositions, etc. etc. By LEROY J. BLINN, Master Mechanic. With over One Hundred Illustrations. 12mo. $2 50

BOOTH.—MARBLE WORKER'S MANUAL:
Containing Practical Information respecting Marbles in general, their Cutting, Working, and Polishing; Veneering of Marble; Mosaics; Composition and Use of Artificial Marble, Stuccos, Cements, Receipts, Secrets, etc. etc. Translated from the French by M. L. BOOTH. With an Appendix concerning American Marbles. 12mo., cloth . . $1 50

BOOTH AND MORFIT.—THE ENCYCLOPEDIA OF CHEMISTRY, PRACTICAL AND THEORETICAL:
Embracing its application to the Arts, Metallurgy, Mineralogy, Geology, Medicine, and Pharmacy. By JAMES C. BOOTH, Melter and Refiner in the United States Mint, Professor of Applied Chemistry in the Franklin Institute, etc., assisted by CAMPBELL MORFIT, author of "Chemical Manipulations," etc. Seventh edition. Complete in one volume, royal 8vo., 978 pages, with numerous wood-cuts and other illustrations. $5 00

BOWDITCH.—ANALYSIS, TECHNICAL VALUATION, PURIFICATION, AND USE OF COAL GAS:
By Rev. W. R. BOWDITCH. Illustrated with wood engravings. 8vo. $6 50

BOX.—PRACTICAL HYDRAULICS:
A Series of Rules and Tables for the use of Engineers, etc. By THOMAS BOX. 12mo. $2 00

BUCKMASTER.—THE ELEMENTS OF MECHANICAL PHYSICS:
By J. C. BUCKMASTER, late Student in the Government School of Mines; Certified Teacher of Science by the Department of Science and Art; Examiner in Chemistry and Physics in the Royal College of Preceptors; and late Lecturer in Chemistry and Physics of the Royal Polytechnic Institute. Illustrated with numerous engravings. In one vol. 12mo. . $2 00

BULLOCK.—THE AMERICAN COTTAGE BUILDER:
A Series of Designs, Plans, and Specifications, from $200 to to $20,000 for Homes for the People; together with Warming, Ventilation, Drainage, Painting, and Landscape Gardening. By JOHN BULLOCK, Architect, Civil Engineer, Mechanician, and Editor of "The Rudiments of Architecture and Building," etc. Illustrated by 75 engravings. In one vol. 8vo. $3 50

BULLOCK.—THE RUDIMENTS OF ARCHITECTURE AND BUILDING:

For the use of Architects, Builders, Draughtsmen, Machinists, Engineers, and Mechanics. Edited by JOHN BULLOCK, author of "The American Cottage Builder." Illustrated by 250 engravings. In one volume 8vo. . . . $3 50

BURGH.—PRACTICAL ILLUSTRATIONS OF LAND AND MARINE ENGINES:

Showing in detail the Modern Improvements of High and Low Pressure, Surface Condensation, and Super-heating, together with Land and Marine Boilers. By N. P. BURGH, Engineer. Illustrated by twenty plates, double elephant folio, with text. $21 00

BURGH.—PRACTICAL RULES FOR THE PROPORTIONS OF MODERN ENGINES AND BOILERS FOR LAND AND MARINE PURPOSES.

By N. P. BURGH, Engineer. 12mo. . . . $2 00

BURGH.—THE SLIDE-VALVE PRACTICALLY CONSIDERED:

By N. P. BURGH, author of "A Treatise on Sugar Machinery," "Practical Illustrations of Land and Marine Engines," "A Pocket-Book of Practical Rules for Designing Land and Marine Engines, Boilers," etc. etc. etc. Completely illustrated. 12mo. $2 00

BYRN.—THE COMPLETE PRACTICAL BREWER:

Or, Plain, Accurate, and Thorough Instructions in the Art of Brewing Beer, Ale, Porter, including the Process of making Bavarian Beer, all the Small Beers, such as Root-beer, Ginger-pop, Sarsaparilla-beer, Mead, Spruce beer, etc. etc. Adapted to the use of Public Brewers and Private Families. By M. LA FAYETTE BYRN, M. D. With illustrations. 12mo. $1 25

BYRN.—THE COMPLETE PRACTICAL DISTILLER:

Comprising the most perfect and exact Theoretical and Practical Description of the Art of Distillation and Rectification; including all of the most recent improvements in distilling apparatus; instructions for preparing spirits from the numerous vegetables, fruits, etc.; directions for the distillation and preparation of all kinds of brandies and other spirits, spirituous and other compounds, etc. etc.; all of which is so simplified that it is adapted not only to the use of extensive distillers, but for every farmer, or others who may wish to engage in the art of distilling By M. LA FAYETTE BYRN, M. D. With numerous engravings. In one volume, 12mo. $1 50

BYRNE.—POCKET BOOK FOR RAILROAD AND CIVIL ENGINEERS:

Containing New, Exact, and Concise Methods for Laying out Railroad Curves, Switches, Frog Angles and Crossings; the Staking out of work; Levelling; the Calculation of Cuttings; Embankments; Earth-work, etc. By OLIVER BYRNE. Illustrated, 18mo. $1 25

BYRNE.—THE HANDBOOK FOR THE ARTISAN, MECHANIC, AND ENGINEER:

By OLIVER BYRNE. Illustrated by 11 large plates and 185 Wood Engravings. 8vo. $5 00

BYRNE.—THE ESSENTIAL ELEMENTS OF PRACTICAL MECHANICS:

For Engineering Students, based on the Principle of Work. By OLIVER BYRNE. Illustrated by Numerous Wood Engravings, 12mo. $3 63

BYRNE.—THE PRACTICAL METAL-WORKER'S ASSISTANT:

Comprising Metallurgic Chemistry; the Arts of Working all Metals and Alloys; Forging of Iron and Steel; Hardening and Tempering; Melting and Mixing; Casting and Founding; Works in Sheet Metal; the Processes Dependent on the Ductility of the Metals; Soldering; and the most Improved Processes and Tools employed by Metal-Workers. With the Application of the Art of Electro-Metallurgy to Manufacturing Processes; collected from Original Sources, and from the Works of Holtzapffel, Bergeron, Leupold, Plumier, Napier, and others. By OLIVER BYRNE. A New, Revised, and improved Edition, with Additions by John Scoffern, M. B , William Clay, Wm. Fairbairn, F. R. S., and James Napier. With Five Hundred and Ninety-two Engravings; Illustrating every Branch of the Subject. In one volume, 8vo. 652 pages . $7 00

BYRNE.—THE PRACTICAL CALCULATOR:

For the Engineer, Mechanic, Manufacturer of Engine Work, Naval Architect, Miner, and Millwright. By OLIVER BYRNE. 1 volume, 8vo., nearly 600 pages $4 50

CABINET MAKER'S ALBUM OF FURNITURE:

Comprising a Collection of Designs for the Newest and Most Elegant Styles of Furniture. Illustrated by Forty eight Large and Beautifully Engraved Plates. In one volume, oblong $5 00

CALVERT.—LECTURES ON COAL-TAR COLORS, AND ON RECENT IMPROVEMENTS AND PROGRESS IN DYEING AND CALICO PRINTING:

Embodying Copious Notes taken at the last London International Exhibition, and *Illustrated with Numerous Patterns of Aniline and other Colors.* By F. GRACE CALVERT, F. R. S., F. C. S., Professor of Chemistry at the Royal Institution, Manchester, Corresponding Member of the Royal Academies of Turin and Rouen; of the Pharmaceutical Society of Paris; Société Industrielle de Mulhouse, etc. In one volume, 8vo., cloth $1 50

CAMPIN.—A PRACTICAL TREATISE ON MECHANICAL ENGINEERING:

Comprising Metallurgy, Moulding, Casting, Forging, Tools, Workshop Machinery, Mechanical Manipulation, Manufacture of Steam-engines, etc. etc. With an Appendix on the Analysis of Iron and Iron Ores. By FRANCIS CAMPIN, C. E. To which are added, Observations on the Construction of Steam Boilers, and Remarks upon Furnaces used for Smoke Prevention; with a Chapter on Explosions. By R. Armstrong, C. E., and John Bourne. Rules for Calculating the Change Wheels for Screws on a Turning Lathe, and for a Wheel-cutting Machine. By J. LA NICCA. Management of Steel, including Forging, Hardening, Tempering, Annealing, Shrinking, and Expansion. And the Case-hardening of Iron. By G. EDE. 8vo. Illustrated with 29 plates and 100 wood engravings. $6 00

CAMPIN.—THE PRACTICE OF HAND-TURNING IN WOOD, IVORY, SHELL, ETC.:

With Instructions for Turning such works in Metal as may be required in the Practice of Turning Wood, Ivory, etc. Also, an Appendix on Ornamental Turning. By FRANCIS CAMPIN; with Numerous Illustrations, 12mo., cloth . . . $3 00

CAPRON DE DOLE.—DUSSAUCE.—BLUES AND CARMINES OF INDIGO.

A Practical Treatise on the Fabrication of every Commercial Product derived from Indigo. By FELICIEN CAPRON DE DOLE. Translated, with important additions, by Professor H. DUSSAUCE. 12mo. $2 50

CAREY.—THE WORKS OF HENRY C. CAREY:

CONTRACTION OR EXPANSION? REPUDIATION OR RESUMPTION? Letters to Hon. Hugh McCulloch. 8vo. 38

FINANCIAL CRISES, their Causes and Effects. 8vo. paper 25

HARMONY OF INTERESTS; Agricultural, Manufacturing, and Commercial. 8vo., paper $1 00
Do. do. cloth $1 50

LETTERS TO THE PRESIDENT OF THE UNITED STATES. Paper 75

MANUAL OF SOCIAL SCIENCE. Condensed from Carey's "Principles of Social Science." By KATE MCKEAN. 1 vol. 12mo. $2 25

MISCELLANEOUS WORKS: comprising "Harmony of Interests," "Money," "Letters to the President," "French and American Tariffs," "Financial Crises," "The Way to Outdo England without Fighting Her," "Resources of the Union," "The Public Debt," "Contraction or Expansion," "Review of the Decade 1857—'67," "Reconstruction," etc. etc. 1 vol. 8vo., cloth $4 50

MONEY: A LECTURE before the N. Y. Geographical and Statistical Society. 8vo., paper 25

PAST, PRESENT, AND FUTURE. 8vo. $2 50

PRINCIPLES OF SOCIAL SCIENCE. 3 volumes 8vo., cloth $10 00

REVIEW OF THE DECADE 1857—'67. 8vo., paper 38

RECONSTRUCTION: INDUSTRIAL, FINANCIAL, AND POLITICAL. Letters to the Hon. Henry Wilson, U. S. S. 8vo. paper 38

THE PUBLIC DEBT, LOCAL AND NATIONAL. How to provide for its discharge while lessening the burden of Taxation. Letter to David A. Wells, Esq., U. S. Revenue Commission. 8vo., paper 25

THE RESOURCES OF THE UNION. A Lecture read, Dec. 1865, before the American Geographical and Statistical Society, N. Y., and before the American Association for the Advancement of Social Science, Boston 25

THE SLAVE TRADE, DOMESTIC AND FOREIGN; Why it Exists, and How it may be Extinguished. 12mo., cloth $1 50

THE WAY TO OUTDO ENGLAND WITHOUT FIGHTING HER. Letters to the Hon. Schuyler Colfax, Speaker of the House of Representatives United States, on "The Paper Question," "The Farmer's Question," "The Iron Question," "The Railroad Question," and "The Currency Question." 8vo., paper 75

CHEVALIER.—THE PHOTOGRAPHIC STUDENT.

A Complete Treatise on the Theory and Practice of Photography. Translated from the French of A. CHEVALIER. Illustrated by numerous engravings. (In press.)

CLOUGH.—THE CONTRACTOR'S MANUAL AND BUILDER'S PRICE-BOOK:

Designed to elucidate the method of ascertaining, correctly, the value and Quantity of every description of Work and Materials used in the Art of Building, from their Prime Cost in any part of the United States, collected from extensive experience and observation in Building and Designing; to which are added a large variety of Tables, Memoranda, etc., indispensable to all engaged or concerned in erecting buildings of any kind. By A. B. CLOUGH, Architect, 24mo., cloth 75

COLBURN.—THE GAS-WORKS OF LONDON:

Comprising a sketch of the Gas-works of the city, Process of Manufacture, Quantity Produced, Cost, Profit, etc. By ZERAH COLBURN. 8vo., cloth 75

COLBURN.—THE LOCOMOTIVE ENGINE:

Including a Description of its Structure, Rules for Estimating its Capabilities, and Practical Observations on its Construction and Management. By ZERAH COLBURN. Illustrated. A new edition. 12mo. $1 25

COLBURN AND MAW.—THE WATER-WORKS OF LONDON:

Together with a Series of Articles on various other Waterworks. By ZERAH COLBURN and W. MAW. Reprinted from "Engineering." In one volume, 8vo. . . $4 00

DAGUERREOTYPIST AND PHOTOGRAPHER'S COMPANION:

12mo., cloth $1 25

DAVIS.—A TREATISE ON HARNESS, SADDLES, AND BRIDLES:

Their History and Manufacture from the Earliest Times down to the Present Period. By A. DAVIS, Practical Saddler and Harness Maker. (In press.)

DESSOYE.—STEEL, ITS MANUFACTURE, PROPERTIES, AND USE.

By J. B. J. Dessoye, Manufacturer of Steel; with an Introduction and Notes by Ed. Graten, Engineer of Mines. Translated from the French. In one volume, 12mo. (In press.)

DIRCKS.—PERPETUAL MOTION:

Or Search for Self-Motive Power during the 17th, 18th, and 19th centuries. Illustrated from various authentic sources in Papers, Essays, Letters, Paragraphs, and numerous Patent Specifications, with an Introductory Essay by Henry Dircks, C. E. Illustrated by numerous engravings of machines. 12mo., cloth $3 50

DIXON.—THE PRACTICAL MILLWRIGHT'S AND ENGINEER'S GUIDE:

Or Tables for Finding the Diameter and Power of Cogwheels; Diameter, Weight, and Power of Shafts; Diameter and Strength of Bolts, etc. etc. By Thomas Dixon. 12mo., cloth. $1 50

DUNCAN.—PRACTICAL SURVEYOR'S GUIDE:

Containing the necessary information to make any person, of common capacity, a finished land surveyor without the aid of a teacher. By Andrew Duncan. Illustrated. 12mo., cloth. $1 25

DUSSAUCE.—A NEW AND COMPLETE TREATISE ON THE ARTS OF TANNING, CURRYING, AND LEATHER DRESSING:

Comprising all the Discoveries and Improvements made in France, Great Britain, and the United States. Edited from Notes and Documents of Messrs. Sallerou, Grouvelle, Duval, Dessables, Labarraque, Payen, René, De Fontenelle, Malapeyre, etc. etc. By Prof. H. Dussauce, Chemist. Illustrated by 212 wood engravings. 8vo. $10 00

DUSSAUCE.—A GENERAL TREATISE ON THE MANUFACTURE OF EVERY DESCRIPTION OF SOAP:

Comprising the Chemistry of the Art, with Remarks on Alkalies, Saponifiable Fatty Bodies, the apparatus necessary in a Soap Factory, Practical Instructions on the manufacture of the various kinds of Soap, the assay of Soaps, etc. etc. Edited from notes of Larmé, Fontenelle, Malapeyre, Dufour, and others, with large and important additions by Professor H. Dussauce, Chemist. Illustrated. In one volume, 8vo. (In press.)

DUSSAUCE.—A PRACTICAL GUIDE FOR THE PERFUMER:
Being a New Treatise on Perfumery the most favorable to the Beauty without being injurious to the Health, comprising a Description of the substances used in Perfumery, the Formulæ of more than one thousand Preparations, such as Cosmetics, Perfumed Oils, Tooth Powders, Waters, Extracts, Tinctures, Infusions, Vinaigres, Essential Oils, Pastels, Creams, Soaps, and many new Hygienic Products not hitherto described. Edited from Notes and Documents of Messrs. Debay, Lunel, etc. With additions by Professor H. Dussauce, Chemist. (In press, *shortly to be issued.*)

DUSSAUCE.—PRACTICAL TREATISE ON THE FABRICATION OF MATCHES, GUN COTTON, AND FULMINATING POWDERS.
By Professor H. Dussauce. 12mo. $3 00

DUSSAUCE.—TREATISE ON THE COLORING MATTERS DERIVED FROM COAL TAR:
Their Practical Application in Dyeing Cotton, Wool, and Silk; the Principles of the Art of Dyeing and of the Distillation of Coal Tar, with a Description of the most Important New Dyes now in use. By Prof. H. Dussauce. 12mo. . $3 00

DYER AND COLOR-MAKER'S COMPANION:
Containing upwards of two hundred Receipts for making Colors, on the most approved principles, for all the various styles and fabrics now in existence; with the Scouring Process, and plain Directions for Preparing, Washing-off, and Finishing the Goods. In one vol. 12mo. $1 25

EASTON.—A PRACTICAL TREATISE ON STREET OR HORSE-POWER RAILWAYS:
Their Location, Construction, and Management; with General Plans and Rules for their Organization and Operation; together with Examinations as to their Comparative Advantages over the Omnibus System, and Inquiries as to their Value for Investment; including Copies of Municipal Ordinances relating thereto. By Alexander Easton, C. E. Illustrated by 23 plates, 8vo., cloth $2 00

ERNI.—COAL OIL AND PETROLEUM:
Their Origin, History, Geology, and Chemistry; with a view of their importance in their bearing on National Industry. By Dr. Henri Erni, Chief Chemist, Department of Agriculture. 12mo. $2 50

ERNI.—THE THEORETICAL AND PRACTICAL CHEMISTRY OF FERMENTATION:

Comprising the Chemistry of Wine, Beer, Distilling of Liquors; with the Practical Methods of their Chemical Examination, Preservation, and Improvement—such as Gallizing of Wines. With an Appendix, containing well-tested Practical Rules and Receipts for the manufacture, etc., of all kinds of Alcoholic Liquors. By HENRY ERNI, Chief Chemist, Department of Agriculture. (In press.)

FAIRBAIRN.—THE PRINCIPLES OF MECHANISM AND MACHINERY OF TRANSMISSION:

Comprising the Principles of Mechanism, Wheels, and Pulleys, Strength and Proportions of Shafts, Couplings of Shafts, and Engaging and Disengaging Gear. By WILLIAM FAIRBAIRN, Esq., C. E., LL. D., F. R. S., F. G. S., Corresponding Member of the National Institute of France, and of the Royal Academy of Turin; Chevalier of the Legion of Honor, etc. etc. Beautifully illustrated by over 150 wood-cuts. In one volume 12mo. $2 50

FAIRBAIRN.—PRIME-MOVERS:

Comprising the Accumulation of Water-power; the Construction of Water-wheels and Turbines; the Properties of Steam; the Varieties of Steam-engines and Boilers and Wind-mills. By WILLIAM FAIRBAIRN, C. E., LL. D., F. R. S., F. G. S. Author of "Principles of Mechanism and the Machinery of Transmission." With Numerous Illustrations. In one volume. (In press.)

FLAMM.—A PRACTICAL GUIDE TO THE CONSTRUCTION OF ECONOMICAL HEATING APPLICATIONS FOR SOLID AND GASEOUS FUELS:

With the Application of Concentrated Heat, and on Waste Heat, for the Use of Engineers, Architects, Stove and Furnace Makers, Manufacturers of Fire Brick, Zinc, Porcelain, Glass, Earthenware, Steel, Chemical Products, Sugar Refiners, Metallurgists, and all others employing Heat. By M. PIERRE FLAMM, Manufacturer. Illustrated. Translated from the French. One volume, 12mo. (In press.)

GILBART.—A PRACTICAL TREATISE ON BANKING:

By JAMES WILLIAM GILBART. To which is added: THE NATIONAL BANK ACT AS NOW (1868) IN FORCE. 8vo. $4 50

GOTHIC ALBUM FOR CABINET MAKERS:
Comprising a Collection of Designs for Gothic Furniture. Illustrated by twenty-three large and beautifully engraved plates. Oblong $3 00

GRANT.—BEET-ROOT SUGAR AND CULTIVATION OF THE BEET:
By E. B. GRANT. 12mo. $1 25

GREGORY.—MATHEMATICS FOR PRACTICAL MEN:
Adapted to the Pursuits of Surveyors, Architects, Mechanics, and Civil Engineers. By OLINTHUS GREGORY. 8vo., plates, cloth $3 00

GRISWOLD.—RAILROAD ENGINEER'S POCKET COMPANION.
Comprising Rules for Calculating Deflection Distances and Angles, Tangential Distances and Angles, and all Necessary Tables for Engineers; also the art of Levelling from Preliminary Survey to the Construction of Railroads, intended Expressly for the Young Engineer, together with Numerous Valuable Rules and Examples. By W. GRISWOLD. 12mo., tucks. $1 25

GUETTIER.—METALLIC ALLOYS:
Being a Practical Guide to their Chemical and Physical Properties, their Preparation, Composition, and Uses. Translated from the French of A. GUETTIER, Engineer and Director of Founderies, author of "La Fouderie en France," etc. etc. By A. A. FESQUET, Chemist and Engineer. In one volume, 12mo. (In press, *shortly to be published.*)

HATS AND FELTING:
A Practical Treatise on their Manufacture. By a Practical Hatter. Illustrated by Drawings of Machinery, &c., 8vo.

HAY.—THE INTERIOR DECORATOR:
The Laws of Harmonious Coloring adapted to Interior Decorations: with a Practical Treatise on House-Painting. By D. R. HAY, House-Painter and Decorator. Illustrated by a Diagram of the Primary, Secondary, and Tertiary Colors. 12mo. $2 25

HUGHES.—AMERICAN MILLER AND MILLWRIGHT'S ASSISTANT:
By WM. CARTER HUGHES. A new edition. In one volume, 12mo. $1 50

HUNT.—THE PRACTICE OF PHOTOGRAPHY.

By Robert Hunt, Vice-President of the Photographic Society, London, with numerous illustrations. 12mo., cloth . 75

HURST.—A HAND-BOOK FOR ARCHITECTURAL SURVEYORS:

Comprising Formulæ useful in Designing Builder's work, Table of Weights, of the materials used in Building, Memoranda connected with Builders' work, Mensuration, the Practice of Builders' Measurement, Contracts of Labor, Valuation of Property, Summary of the Practice in Dilapidation, etc. etc. By J. F. Hurst, C. E. 2d edition, pocket-book form, full bound $2 50

JERVIS.—RAILWAY PROPERTY:

A Treatise on the Construction and Management of Railways; designed to afford useful knowledge, in the popular style, to the holders of this class of property; as well as Railway Managers, Officers, and Agents. By John B. Jervis, late Chief Engineer of the Hudson River Railroad, Croton Aqueduct, &c. One vol. 12mo., cloth $2 00

JOHNSON.—A REPORT TO THE NAVY DEPARTMENT OF THE UNITED STATES ON AMERICAN COALS:

Applicable to Steam Navigation and to other purposes. By Walter R. Johnson. With numerous illustrations. 607 pp. 8vo., half morocco $6 00

JOHNSON.—THE COAL TRADE OF BRITISH AMERICA:

With Researches on the Characters and Practical Values of American and Foreign Coals. By Walter R. Johnson, Civil and Mining Engineer and Chemist. 8vo. $2 00

JOHNSTON.—INSTRUCTIONS FOR THE ANALYSIS OF SOILS, LIMESTONES, AND MANURES.

By J. W. F. Johnston. 12mo. 38

KEENE.—A HAND-BOOK OF PRACTICAL GAUGING,

For the Use of Beginners, to which is added A Chapter on Distillation, describing the process in operation at the Custom House for ascertaining the strength of wines. By James B. Keene, of H. M. Customs. 8vo. $1 25

KENTISH.—A TREATISE ON A BOX OF INSTRUMENTS,

And the Slide Rule; with the Theory of Trigonometry and Logarithms, including Practical Geometry, Surveying, Measuring of Timber, Cask and Malt Gauging, Heights, and Distances. By Thomas Kentish. In one volume. 12mo. . $1 25

KOBELL.—ERNI.—MINERALOGY SIMPLIFIED:

A short method of Determining and Classifying Minerals, by means of simple Chemical Experiments in the Wet Way. Translated from the last German Edition of F. VON KOBELL, with an Introduction to Blowpipe Analysis and other additions. By HENRI ERNI, M. D., Chief Chemist, Department of Agriculture, author of "Coal Oil and Petroleum." In one volume, 12mo. $2 50

LAFFINEUR.—A PRACTICAL GUIDE TO HYDRAULICS FOR TOWN AND COUNTRY;

Or a Complete Treatise on the Building of Conduits for Water for Cities, Towns, Farms, Country Residences, Workshops, etc. Comprising the means necessary for obtaining at all times abundant supplies of Drinkable Water. Translated from the French of M. JULES LAFFINEUR, C. E. Illustrated. (In press.)

LAFFINEUR.—A TREATISE ON THE CONSTRUCTION OF WATER-WHEELS:

Containing the various Systems in use with Practical Information on the Dimensions necessary for Shafts, Journals, Arms, etc., of Water-wheels, etc. etc. Translated from the French of M. JULES LAFFINEUR, C. E. Illustrated by numerous plates. (In press.)

LANDRIN.—A TREATISE ON STEEL:

Comprising the Theory, Metallurgy, Practical Working, Properties, and Use. Translated from the French of H. C. LANDRIN, Jr., C. E. By A. A. FESQUET, Chemist and Engineer. Illustrated. 12mo. (In press.)

LARKIN.—THE PRACTICAL BRASS AND IRON FOUNDER'S GUIDE:

A Concise Treatise on Brass Founding, Moulding, the Metals and their Alloys, etc.; to which are added Recent Improvements in the Manufacture of Iron, Steel by the Bessemer Process, etc. etc. By JAMES LARKIN, late Conductor of the Brass Foundry Department in Reany, Neafie & Co.'s Penn Works, Philadelphia. Fifth edition, revised, with Extensive additions. In one volume, 12mo. $2 25

LEAVITT.—FACTS ABOUT PEAT AS AN ARTICLE OF FUEL:

With Remarks upon its Origin and Composition, the Localities in which it is found, the Methods of Preparation and Manufacture, and the various Uses to which it is applicable; together with many other matters of Practical and Scientific Interest. To which is added a chapter on the Utilization of Coal Dust with Peat for the Production of an Excellent Fuel at Moderate Cost, especially adapted for Steam Service. By H. T. Leavitt. Third edition. 12mo. $1 75

LEROUX.—A PRACTICAL TREATISE ON WOOLS AND WORSTEDS:

Translated from the French of Charles Leroux, Mechanical Engineer, and Superintendent of a Spinning Mill. Illustrated by 12 large plates and 34 engravings. In one volume 8vo. (In press, *shortly to be published.*)

LESLIE (MISS).—COMPLETE COOKERY:

Directions for Cookery in its Various Branches. By Miss Leslie. 58th thousand. Thoroughly revised, with the addition of New Receipts. In 1 vol. 12mo., cloth . . $1 25

LESLIE (MISS). LADIES' HOUSE BOOK:

a Manual of Domestic Economy. 20th revised edition. 12mo., cloth $1 25

LESLIE (MISS).—TWO HUNDRED RECEIPTS IN FRENCH COOKERY.

12mo. 50

LIEBER.—ASSAYER'S GUIDE:

Or, Practical Directions to Assayers, Miners, and Smelters, for the Tests and Assays, by Heat and by Wet Processes, for the Ores of all the principal Metals, of Gold and Silver Coins and Alloys, and of Coal, etc. By Oscar M. Lieber. 12mo., cloth $1 25

LOVE.—THE ART OF DYEING, CLEANING, SCOURING, AND FINISHING:

On the most approved English and French methods; being Practical Instructions in Dyeing Silks, Woollens, and Cottons, Feathers, Chips, Straw, etc.; Scouring and Cleaning Bed and Window Curtains, Carpets, Rugs, etc.; French and English Cleaning, any Color or Fabric of Silk, Satin, or Damask. By Thomas Love, a Working Dyer and Scourer. In 1 vol. 12mo. $3 00

MAIN AND BROWN.—QUESTIONS ON SUBJECTS CONNECTED WITH THE MARINE STEAM-ENGINE:

And Examination Papers; with Hints for their Solution. By THOMAS J. MAIN, Professor of Mathematics, Royal Naval College, and THOMAS BROWN, Chief Engineer, R. N. 12mo., cloth $1 50

MAIN AND BROWN.—THE INDICATOR AND DYNAMOMETER:

With their Practical Applications to the Steam-Engine. By THOMAS J. MAIN, M. A. F. R., Ass't Prof. Royal Naval College, Portsmouth, and THOMAS BROWN, Assoc. Inst. C. E., Chief Engineer, R. N., attached to the R. N. College. Illustrated. From the Fourth London Edition. 8vo. . . . $1 50

MAIN AND BROWN.—THE MARINE STEAM-ENGINE.

By THOMAS J. MAIN, F. R. Ass't S. Mathematical Professor at Royal Naval College, and THOMAS BROWN, Assoc. Inst. C. E. Chief Engineer, R. N. Attached to the Royal Naval College. Authors of "Questions connected with the Marine Steam-Engine," and the "Indicator and Dynamometer." With numerous Illustrations. In one volume, 8vo. . . . $5 00

MAKINS.—A MANUAL OF METALLURGY:

More particularly of the Precious Metals: including the Methods of Assaying them. Illustrated by upwards of 50 Engravings. By GEORGE HOGARTH MAKINS, M. R. C. S., F. C. S., one of the Assayers to the Bank of England, Assayer to the Anglo-Mexican Mints, and Lecturer upon Metallurgy at the Dental Hospital, London. In one volume, 12mo. . . $3 50

MARTIN.—SCREW-CUTTING TABLES, FOR THE USE OF MECHANICAL ENGINEERS:

Showing the Proper Arrangement of Wheels for Cutting the Threads of Screws of any required Pitch; with a Table for Making the Universal Gas-Pipe Thread and Taps. By W. A. MARTIN, Engineer. 8vo. 50

MILES.—A PLAIN TREATISE ON HORSE-SHOEING.

With illustrations. By WILLIAM MILES, author of "The Horse's Foot," $1 00

MOLESWORTH. POCKET-BOOK OF USEFUL FORMULÆ AND MEMORANDA FOR CIVIL AND MECHANICAL ENGINEERS.

By GUILFORD L. MOLESWORTH, Member of the Institution of Civil Engineers, Chief Resident Engineer of the Ceylon Railway. Second American, from the Tenth London Edition. In one volume, full bound in pocket-book form . . $2 00

MOORE.—THE INVENTOR'S GUIDE:

Patent Office and Patent Laws; or, a Guide to Inventors, and a Book of Reference for Judges, Lawyers, Magistrates, and others. By J. G. MOORE. 12mo., cloth . . $1 25

MOREAU.—PRACTICAL GUIDE FOR THE JEWELLER,

In the Application of Harmony of Colors in the Arrangement of Precious Stones, Gold, etc., from the French of M. L. MOREAU, Jeweller and Designer. Illustrated. (In press.)

NAPIER.—CHEMISTRY APPLIED TO DYEING.

By JAMES NAPIER, F. C. S. A new and revised edition, brought down to the present condition of the Art. Illustrated. (In press.)

NAPIER.—A MANUAL OF DYEING RECEIPTS FOR GENERAL USE.

By JAMES NAPIER, F. C. S. *With Numerous Patterns of Dyed Cloth and Silk.* Second edition, revised and enlarged. 12mo. $3 75

NAPIER.—MANUAL OF ELECTRO-METALLURGY:

Including the Application of the Art to Manufacturing Processes. By JAMES NAPIER. Fourth American, from the Fourth London edition, revised and enlarged. Illustrated by engravings. In one volume, 8vo. $2 00

NEWBERY.—GLEANINGS FROM ORNAMENTAL ART OF EVERY STYLE;

Drawn from Examples in the British, South Kensington, Indian, Crystal Palace, and other Museums, the Exhibitions of 1851 and 1862, and the best English and Foreign works. In a series of one hundred exquisitely drawn Plates, containing many hundred examples. By ROBERT NEWBERY. 4to. $15 00

NICHOLSON.—A MANUAL OF THE ART OF BOOK-BINDING:

Containing full instructions in the different Branches of Forwarding, Gilding, and Finishing. Also, the Art of Marbling Book-edges and Paper. By JAMES B. NICHOLSON. Illustrated. 12mo., cloth $2 25

NORRIS.—A HAND-BOOK FOR LOCOMOTIVE ENGINEERS AND MACHINISTS:

Comprising the Proportions and Calculations for Constructing Locomotives; Manner of Setting Valves; Tables of Squares, Cubes, Areas, etc. etc. By SEPTIMUS NORRIS, Civil and Mechanical Engineer. New edition. Illustrated, 12mo., cloth $2 00

NYSTROM.—ON TECHNOLOGICAL EDUCATION AND THE CONSTRUCTION OF SHIPS AND SCREW PROPELLERS:

For Naval and Marine Engineers. By JOHN W. NYSTROM, late Acting Chief Engineer U. S. N. Second edition, revised with additional matter. Illustrated by seven engravings. 12mo. $2 50

O'NEILL.—CHEMISTRY OF CALICO PRINTING, DYEING, AND BLEACHING:

Including Silken, Woollen, and Mixed Goods; Practical and Theoretical. By CHARLES O'NEILL. (In press.)

O'NEILL.—A DICTIONARY OF CALICO PRINTING AND DYEING:

Containing a Brief Account of all the Substances and Processes in Use in the Arts of Printing and Dyeing Textile Fabrics; with Practical Receipts and Scientific Information. By CHARLES O'NEILL, Analytical Chemist, Fellow of the Chemical Society of London, etc. etc. Author of "Chemistry of Calico Printing and Dyeing." 8vo. (In press.)

OVERMAN—OSBORN.—THE MANUFACTURE OF IRON IN ALL ITS BRANCHES:

Including a Practical Description of the various Fuels and their Values, the Nature, Determination and Preparation of the Ore, the Erection and Management of Blast and other Furnaces, the characteristic results of Working by Charcoal, Coke, or Anthracite, the Conversion of the Crude into the various kinds of Wrought Iron, and the Methods adapted to this end. Also, a Description of Forge Hammers, Rolling Mills, Blast Engines, &c. &c. To which is added an Essay on the Manufacture of Steel. By FREDERICK OVERMAN, Mining Engineer. The whole thoroughly revised and enlarged, adapted to the latest Improvements and Discoveries, and the particular type of American Methods of Manufacture. With various new engravings illustrating the whole subject. By H. S. OSBORN, LL. D. Professor of Mining and Metallurgy in Lafayette College. In one volume, 8vo. (In press.) . $10 00

PAINTER, GILDER, AND VARNISHER'S COMPANION:

Containing Rules and Regulations in everything relating to the Arts of Painting, Gilding, Varnishing, and Glass Staining, with numerous useful and valuable Receipts; Tests for the Detection of Adulterations in Oils and Colors, and a statement of the Diseases and Accidents to which Painters, Gilders, and

Varnishers are particularly liable, with the simplest methods of Prevention and Remedy. With Directions for Graining. Marbling, Sign Writing, and Gilding on Glass. To which are added COMPLETE INSTRUCTIONS FOR COACH PAINTING AND VARNISHING. 12mo., cloth $1 50

PALLETT.—THE MILLER'S, MILLWRIGHT'S, AND ENGINEER'S GUIDE.

By HENRY PALLETT. Illustrated. In one vol. 12mo. $3 00

PERKINS.—GAS AND VENTILATION.

Practical Treatise on Gas and Ventilation. With Special Relation to Illuminating, Heating, and Cooking by Gas. Including Scientific Helps to Engineer-students and others. With illustrated Diagrams. By E. E. PERKINS. 12mo., cloth $1 25

PERKINS AND STOWE.—A NEW GUIDE TO THE SHEET-IRON AND BOILER PLATE ROLLER:

Containing a Series of Tables showing the Weight of Slabs and Piles to Produce Boiler Plates, and of the Weight of Piles and the Sizes of Bars to produce Sheet-iron; the Thickness of the Bar Gauge in Decimals; the Weight per foot, and the Thickness on the Bar or Wire Gauge of the fractional parts of an inch; the Weight per sheet, and the Thickness on the Wire Gauge of Sheet-iron of various dimensions to weigh 112 lbs. per bundle; and the conversion of Short Weight into Long Weight, and Long Weight into Short. Estimated and collected by G. H. PERKINS and J. G. STOWE $2 50

PHILLIPS AND DARLINGTON.—RECORDS OF MINING AND METALLURGY:

Or Facts and Memoranda for the use of the Mine Agent and Smelter. By J. ARTHUR PHILLIPS, Mining Engineer, Graduate of the Imperial School of Mines, France, etc., and JOHN DARLINGTON. Illustrated by numerous engravings. In one volume, 12mo. $2 00

PRADAL, MALEPEYRE, AND DUSSAUCE. — A COMPLETE TREATISE ON PERFUMERY:

Containing notices of the Raw Material used in the Art, and the Best Formulæ. According to the most approved Methods followed in France, England, and the United States. By M. P. PRADAL, Perfumer Chemist, and M. F. MALEPEYRE. Translated from the French, with extensive additions, by Professor H. DUSSAUCE. 8vo. $10 00

PROTEAUX.—PRACTICAL GUIDE FOR THE MANUFACTURE OF PAPER AND BOARDS.

By A. Proteaux, Civil Engineer, and Graduate of the School of Arts and Manufactures, Director of Thiers's Paper Mill, 'Puy-de-Dômé. With additions, by L. S. Le Normand. Translated from the French, with Notes, by Horatio Paine, A. B., M. D. To which is added a Chapter on the Manufacture of Paper from Wood in the United States, by Henry T. Brown, of the "American Artisan." Illustrated by six plates, containing Drawings of Raw Materials, Machinery, Plans of Paper-Mills, etc. etc. 8vo. $5 00

REGNAULT.—ELEMENTS OF CHEMISTRY.

By M. V. Regnault. Translated from the French by T. Forrest Betton, M. D., and edited, with notes, by James C. Booth, Melter and Refiner U. S. Mint, and Wm. L. Faber, Metallurgist and Mining Engineer. Illustrated by nearly 700 wood engravings. Comprising nearly 1500 pages. In two volumes, 8vo., cloth $10 00

SELLERS.—THE COLOR MIXER:

Containing nearly Four Hundred Receipts for Colors, Pastes, Acids, Pulps, Blue Vats, Liquors, etc. etc., for Cotton and Woollen Goods: including the celebrated Barrow Delaine Colors. By John Sellers, an experienced Practical Workman. In one volume, 12mo. $2 50

SHUNK.—A PRACTICAL TREATISE ON RAILWAY CURVES AND LOCATION, FOR YOUNG ENGINEERS.

By Wm. F. Shunk, Civil Engineer. 12mo. . . . $1 50

SMEATON.—BUILDER'S POCKET COMPANION:

Containing the Elements of Building, Surveying, and Architecture; with Practical Rules and Instructions connected with the subject. By A. C. Smeaton, Civil Engineer, etc. In one volume, 12mo. $1 25

SMITH.—THE DYER'S INSTRUCTOR:

Comprising Practical Instructions in the Art of Dyeing Silk, Cotton, Wool, and Worsted, and Woollen Goods: containing nearly 800 Receipts. To which is added a Treatise on the Art of Padding; and the Printing of Silk Warps, Skeins, and Handkerchiefs, and the various Mordants and Colors for the different styles of such work. By David Smith, Pattern Dyer. 12mo., cloth. $3 00

SMITH.—PARKS AND PLEASURE GROUNDS:

Or Practical Notes on Country Residences, Villas, Public Parks, and Gardens. By CHARLES H. J. SMITH, Landscape Gardener and Garden Architect, etc. etc. 12mo. . $2 25

STOKES.—CABINET-MAKER'S AND UPHOLSTERER'S COMPANION:

Comprising the Rudiments and Principles of Cabinet-making and Upholstery, with Familiar Instructions, Illustrated by Examples for attaining a Proficiency in the Art of Drawing, as applicable to Cabinet-work; The Processes of Veneering, Inlaying, and Buhl-work; the Art of Dyeing and Staining Wood, Bone, Tortoise Shell, etc. Directions for Lackering, Japanning, and Varnishing; to make French Polish; to prepare the Best Glues, Cements, and Compositions, and a number of Receipts particularly for workmen generally. By J. STOKES. In one vol. 12mo. With illustrations $1 25

STRENGTH AND OTHER PROPERTIES OF METALS.

Reports of Experiments on the Strength and other Properties of Metals for Cannon. With a Description of the Machines for Testing Metals, and of the Classification of Cannon in service. By Officers of the Ordnance Department U. S. Army. By authority of the Secretary of War. Illustrated by 25 large steel plates. In 1 vol. quarto $10 00

TABLES SHOWING THE WEIGHT OF ROUND, SQUARE, AND FLAT BAR IRON, STEEL, ETC.,

By Measurement. Cloth 63

TAYLOR.—STATISTICS OF COAL:

Including Mineral Bituminous Substances employed in Arts and Manufactures; with their Geographical, Geological, and Commercial Distribution and amount of Production and Consumption on the American Continent. With Incidental Statistics of the Iron Manufacture. By R. C. TAYLOR. Second edition, revised by S. S. HALDEMAN. Illustrated by five Maps and many wood engravings. 8vo., cloth . . . $6 00

TEMPLETON.—THE PRACTICAL EXAMINATOR ON STEAM AND THE STEAM-ENGINE:

With Instructive References relative thereto, for the Use of Engineers, Students, and others. By WM. TEMPLETON, Engineer. 12mo. $1 25

THOMAS.—THE MODERN PRACTICE OF PHOTOGRAPHY.
By R. W. Thomas, F. C. S. 8vo., cloth . . . 75

THOMSON.—FREIGHT CHARGES CALCULATOR.
By Andrew Thomson, Freight Agent . . . $1 25

TURNBULL.—THE ELECTRO-MAGNETIC TELEGRAPH:
With an Historical Account of its Rise, Progress, and Present Condition. Also, Practical Suggestions in regard to Insulation and Protection from the effects of Lightning. Together with an Appendix, containing several important Telegraphic Devices and Laws. By Lawrence Tnrnbull, M. D., Lecturer on Technical Chemistry at the Franklin Institute. Revised and improved. Illustrated. 8vo. . . . $3 00

TURNER'S (THE) COMPANION:
Containing Instructions in Concentric, Elliptic, and Eccentric Turning; also various Plates of Chucks, Tools, and Instruments; and Directions for using the Eccentric Cutter, Drill, Vertical Cutter, and Circular Rest; with Patterns and Instructions for working them. A new edition in one vol. 12mo. $1 50

ULRICH—DUSSAUCE.—A COMPLETE TREATISE ON THE ART OF DYEING COTTON AND WOOL:
As practised in Paris, Rouen, Mulhausen, and Germany. From the French of M. Louis Ulrich, a Practical Dyer in the principal Manufactories of Paris, Rouen, Mulhausen, etc. etc.; to which are added the most important Receipts for Dyeing Wool, as practised in the Manufacture Impériale des Gobelins, Paris. By Professor H. Dussauce. 12mo. $3 00

URBIN—BRULL.—A PRACTICAL GUIDE FOR PUDDLING IRON AND STEEL.
By Ed. Urbin, Engineer of Arts and Manufactures. A Prize Essay read before the Association of Engineers, Graduate of the School of Mines, of Liege, Belgium, at the Meeting of 1865—6. To which is added a Comparison of the Resisting Properties of Iron and Steel. By A. Brull. Translated from the French by A. A. Fesquet, Chemist and Engineer. In one volume, 8vo. $1 00

WATSON.—A MANUAL OF THE HAND-LATHE.
By Egbert P. Watson, Late of the "Scientific American," Author of "Modern Practice of American Machinists and Engineers." In one volume, 12mo. (In press.)

WATSON.—THE MODERN PRACTICE OF AMERICAN MACHINISTS AND ENGINEERS:

Including the Construction, Application, and Use of Drills, Lathe Tools, Cutters for Boring Cylinders, and Hollow Work Generally, with the most Economical Speed of the same, the Results verified by Actual Practice at the Lathe, the Vice, and on the Floor. Together with Workshop management, Economy of Manufacture, the Steam-Engine, Boilers, Gears, Belting, etc. etc. By EGBERT P. WATSON, late of the "Scientific American." Illustrated by eighty-six engravings. 12mo. . . $2 50

WATSON.—THE THEORY AND PRACTICE OF THE ART OF WEAVING BY HAND AND POWER:

With Calculations and Tables for the use of those connected with the Trade. By JOHN WATSON, Manufacturer and Practical Machine Maker. Illustrated by large drawings of the best Power-Looms. 8vo. $7 50

WEATHERLY.—TREATISE ON THE ART OF BOILING SUGAR, CRYSTALLIZING, LOZENGE-MAKING, COMFITS, GUM GOODS,

And other processes for Confectionery, &c. In which are explained, in an easy and familiar manner, the various Methods of Manufacturing every description of Raw and Refined sugar Goods, as sold by Confectioners and others . . $2 00

WILL.—TABLES FOR QUALITATIVE CHEMICAL ANALYSIS.

By Prof. HEINRICH WILL, of Giessen, Germany. Seventh edition. Translated by CHARLES F. HIMES, Ph. D., Professor of Natural Science, Dickinson College, Carlisle, Pa. . $1 25

WILLIAMS.—ON HEAT AND STEAM:

Embracing New Views of Vaporization, Condensation, and Expansion. By CHARLES WYE WILLIAMS, A. I. C. E. Illustrated. 8vo. $3 50

www.ingramcontent.com/pod-product-compliance
Lightning Source LLC
LaVergne TN
LVHW010138110826
845151LV00002B/378

* 9 7 8 1 4 2 5 5 3 9 0 8 5 *